Engineering Design Modeling

with

Unigraphics®

Second Edition

Robert A. Chalou

Michigan State University

KENDALL/HUNT PUBLISHING COMPANY
4050 Westmark Drive Dubuque, Iowa 52002

Book Team

Chairman and Chief Executive Officer Mark C. Falb
Vice President, Director of National Book Program Alfred C. Grisanti
Editorial Development Supervisor Georgia Botsford
Developmental Editor Angela Willenbring
Vice President, Production Services Ruth A. Burlage
Production Services Manager Jo Wiegand
Prepress Editor Victoria D. Noel
Design Manager Jodi Splinter
Senior Vice President, College Division Thomas W. Gantz
Vice President and National Field Manager Brian Johnson
Managing Editor, College Field Paul Carty
Associate Editor, College Field David Mattaliano

ISBN 0-7872-9102-1

Printed in the United States of America
10 9 8 7 6 5 4 3 2 1

Foreword

The purpose of this book is to give students a sound, basic knowledge of 3-D Solid Modeling and Assemblies using Unigraphics software. This book is by no means a complete operator's manual for the software. Instead, it is intended to give students the tools to explore the software and the courage to try new functionality, as they become more versiatile designers. Students are encouraged to learn 3-D Solid Modeling early in their educational careers. Learning it early tends to give students a better ability to experiment with shape, dimensions and design intent. Thereby, they are much freer to be creative and can concentrate on the design of the product rather that the final dimensions.

I would like to thank the Partners for the Advancement of CAD/CAM/CAE Education (PACE) - i.e., UGS, General Motors, Electronic Data Systems, and Sun Microsystems - for providing the hardware, software, and training that has made possible 3-D Solid Modeling at my institution, and, therefore, the existance of this book.

I would especially like to thank Hulas King of UGS, for supplying me with the motivation and support to accomplish this task; UGS for supplying me with all screen captures used in this book. I appreciated Dr. Vassilios Theodoracatos' contribution of the puzzle graphic used on the cover, depicting the idea of Unigraphics integration within the engineering curriculum.

Table of Contents

1

Getting Started with Unigraphics

Overview

This chapter begins the modeling of an office chair. We will begin with some of the simpler components and advance to more difficult ones. We will then put all of these components together to create an assembly of the chair. The chair is illustrated below.

In this chapter we will learn how to:

- Start Unigraphics
- Create new and open existing files
- Create components using primitives and features
- Edit features
- Save and close Unigraphics
- Create form features
- Rectangular and circular instance arrays
- Edge blends
- Sketch
- Threading
- Mirror sketches
- Use the WCS, work coordinate system
- Use layers
- Use the model navigator, MNT

1.1.1 Unigraphics Background

Unigraphics had its beginnings with true 3-D modeling in September, 1980 when McAuto released version D2. After many name changes and company reorginazations, Unigraphics has grown into a very sophicated 3-D solid modeling package. Unigraphics is a math based solid modeling tool that uses the Parasolid modeling kernel. It uses the Constructive Solid Geometry, CSG, method to reproduce components. This feature based method uses feature geometry to define the shape of the object instead of just a collection of edges and faces. Features can be easily updated, or edited, simply be redefining the feature geometry. Components made this way represent common manufacturing practices that makes Unigraphics a very powerful tool for engineering design and modeling.

Unigraphics can run with both UNIX and Windows NT operating systems. The interface differences between the two operating systems are very minimal and basically unnoticeable.

This book is intended to get the new user up and running in a short amount of time. We will start with basic commands that we will use quite often and also start building a strong foundation for advanced courses. This book is intended to teach only a small portion of the functionality of Unigraphics, as this is a fully integrated Computer Aided Design, CAD, Computer Aided Manufacturing, CAM, Computer Aided Engineering, CAE, Computer Aided Industrial Design, CAID, and Computer Aided Process Engineering, CAPE, tool. With what we will learn we will in fact, be able to model some very complex components and then assemble these components into assemblies.

Like all programs, Unigraphics stores information in files. Unigraphics modeling files have a .prt file extension. This .prt extension is used for both component modeling and assembly creation.

1.1.2 Starting Unigraphics

The starting procedure for Unigraphics depends on the operating system used at your institution. Additionally for UNIX based systems, it depends on how your institutions system administrator has configured your environment variables. Your instructor will inform you of the starting procedures for UNIX systems. For Windows, you can start Unigraphics with the Start|Programs|Unigraphics V number|Unigraphics sequence. If an icon is installed on your desktop, you can of course double click on it to launch the application.Unigraphics can be pull down menu of icon driven and is completely customizable. We will concentrate on using the pull down menus for most of our work in this book. Your instructor will give you specific instruction on how to customize your interface to suit your needs and tastes.

1.1.3 Applications

Unigraphics has several applications. for this course we will be interested in Gateway, Modeling, and Assemblies. Each application has different tools assigned to it to perform different functions. When application is chosen, a menu is displayed that lists the different applications available.

1.1.4 Gateway

Gateway is the starting point for everything you do with Unigraphics. Because of this, you can do very little while in Gateway. You can open and create files, view existing files, etc. In a moment we will create two new files called seed files. These files will simplify part creation later when we start modeling components for our chair.

The Unigraphics environment has a series of windows. The first set of windows that you will see after starting Unigraphics is the main window and a graphics window as illustrated in figure 1.2. The main window has the pull down tool bars, a dialog box area, menu bar, status line, the work layer toolbar, and the cue line. It is necessary for these to be visible to make working with Unigraphics more productive.

The title bar will inform you which application you are currently in. Some functions are grayed out in gateway because they are not available. Until we get into assemblies, we will be using application modeling after a file is either opened or created. The title bar in the graphics window, lists which file is currently open.

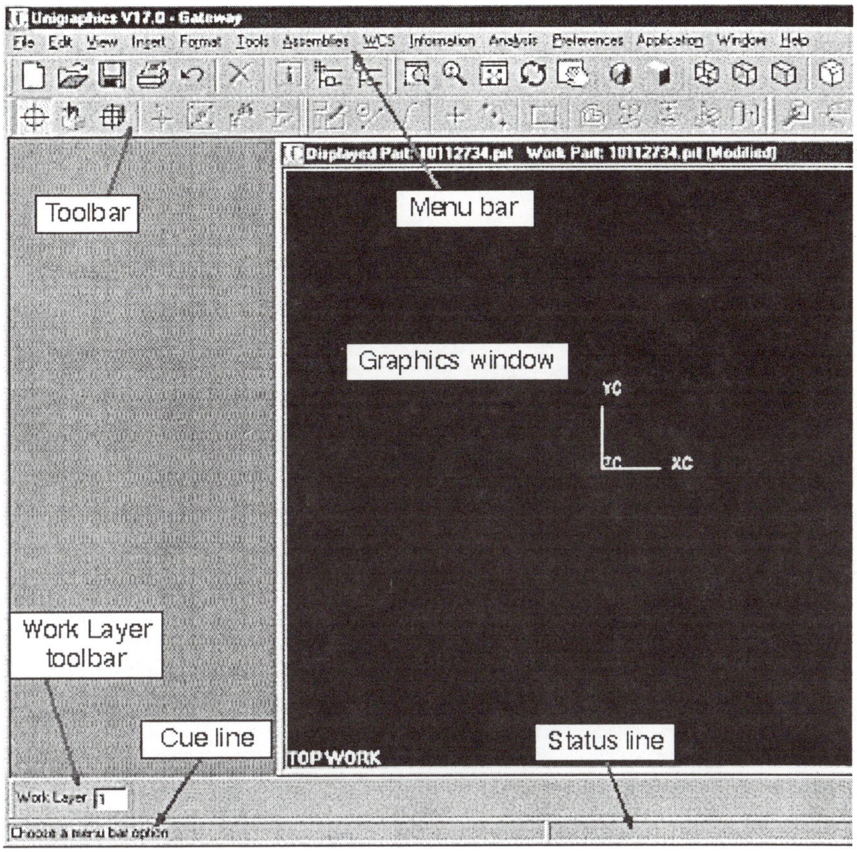

1.1.5 Opening and Creating Files

There are two ways to create a new file. You can use the menu bar with File|New, or the New icon.

You have the opportunity to change measurement units when you create a new file. Be sure to check the units.

Lets now create two part files that we will use over and over again. They are called seed files. Seed files have certain attributes assigned to then that we will alter as we progress through this book. The first will be called seed_in, this part will be the seed part for all our inch parts. The next one is seed_mm, this will be our metric seed part file. Be sure to change units before you say OK.

If you are reopening a part that already exists, you can either use the Open icon, which is next to the New icon or the menu bar with File|Open. The path for this operation will be supplied to you by your instructor.The NT version is shown below and the UNIX version is shown after.

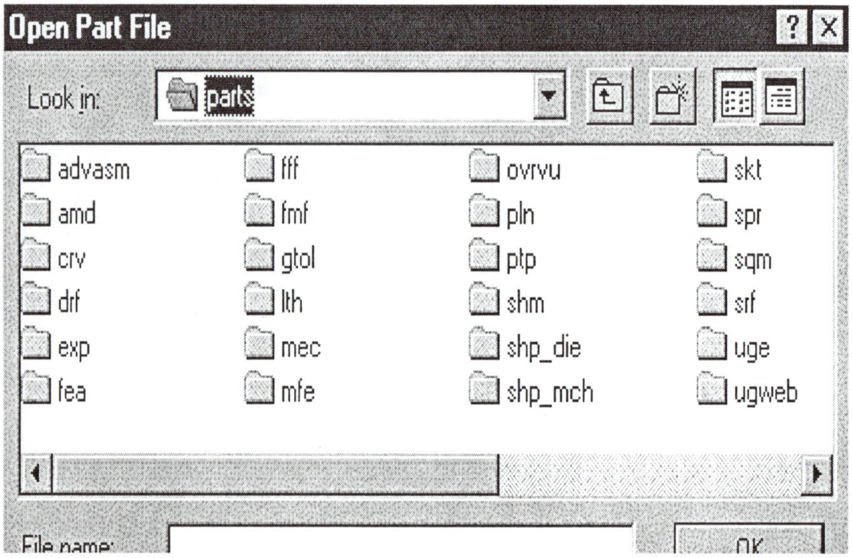

If you open a part that is read only you will get a warning message stating that you will not be able to save your work to that file. Just Ok the message we will learn how to save our work soon.

When you create a new part file, the name that you entered appears on the top of the graphics window, the name of the work view is displayed, (TOP WORK), and the work coordinate system is in the center of the view.

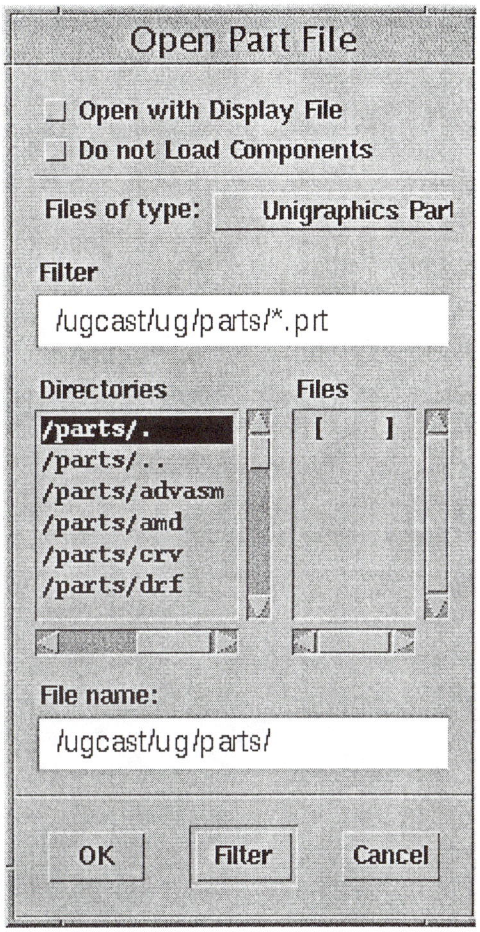

1.1.6 Saving your work

When you open an existing part file you are really working on a copy of the part. This is done on purpose so you can maintain and have design intent. In Unigraphics there is no automatic save. When you create a new part or make changes to an existing part all modifications will be lost unless you save. Establish a routine to save your work when you get to predetermined points.

Again you can save your work two ways. With the save icon or the File|Save pull down menu. Earlier we talked about the read only warning. If you get this, and you want to save the work, you must save your work to a directory that you have right access. In this case you will have to do a save-as command, which is available only from the pull down menu. This will allow you to save a file with either a different name or to a different place or both.

Whenever you see Modified in the graphics window title bar, you must save your work if you want to retain the changes that you made to the component.

1.1.7 Exiting a Unigraphics Session

Choose File | Exit. If you have a file open that is not modified you will be asked if you really want to exit. If you do you would choose OK. If you had a file open that you have made changes on but did not save your work yet, you will get a message telling you that you have modified parts, (parts that have not been saved yet).

1.2 The Chair Base

1.2.1 Lets Begin Modeling

If you haven't already, start a Unigraphics session, open seed_in.prt, and choose application Modeling. This component will be saved as chair_base.prt.

We will start with some easy components. We will be modeling in 3-D so lets change the view to TFR-TRI which is the same as Trimetric. We do this by holding down the right mouse button in the graphics area, choose Replace View, choose TFR-TRI.

For the purposes of this book we will refer to the pull down menus for access to functions in Unigraphics. You notice the work coordinate system move to see all three directions. Right now the work coordinate system, WCS, is at the same place as the absolute coordinate system, ACS. As the name implies the ACS cannot be changed but the WCS moves as needed depending what operation is being performed.

We will start with the base of the chair.

Displayed Part: chair_base.prt   Work Pa

1.2.2 Form Features

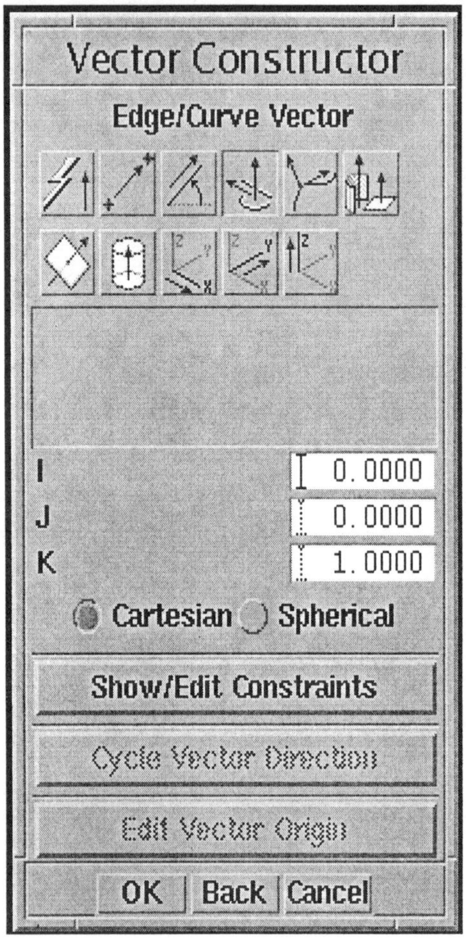

As you can see there are five arms around a central pedestal. Lets start with the pedestal. It is a cylinder that is 3" in diameter by 5" tall. Choose Insert|Form Feature|Cylinder. Notice that the Cue line asks you to choose the creation method. The default method, Diameter, Height, is highlighted, select OK, (the middle mouse button is the same as OK). This will create a cylinder positioned along a vector and will start at a specified origin point which is relative to the WCS. The vector constructor displays asking to chose the cylinder direction.

Lets choose the positive z direction. You are then prompted for the Cylinder parameters. Enter the parameters. You are then prompted to place the origin of the cylinder. Put the center of the bottom face of the cylinder at 0,0,0. If this is not the values that are displayed, choose reset. Choose OK, and a cylinder is created. Cancel the Vector Constructor dialog, we only want one cylinder. Hold the right mouse button in the graphics and fit the view.

1.2.3 Layers

A word about layers. Unigraphics has 256 different layers that act as transparent sheets the are stacked on top of each other. Layers help control and maintain the vast amount of information that is included in the model. Standards are imposed so users can easily fine information in other users files. Your instructor will inform you of the standards that are used at your institution.

Layers are controlled with Format|Layer Settings. You can set layers to Work, Selectable, Invisible and Visible only. One layer has to be the work layer at all times. You can turn off information from view by selecting invisible for one of more layers and the information on that layer will be hidden from view. Selectable means that the information on that layer is visible and you can work with the information on that layer. Visible means that you can see the information but it is not selectable.

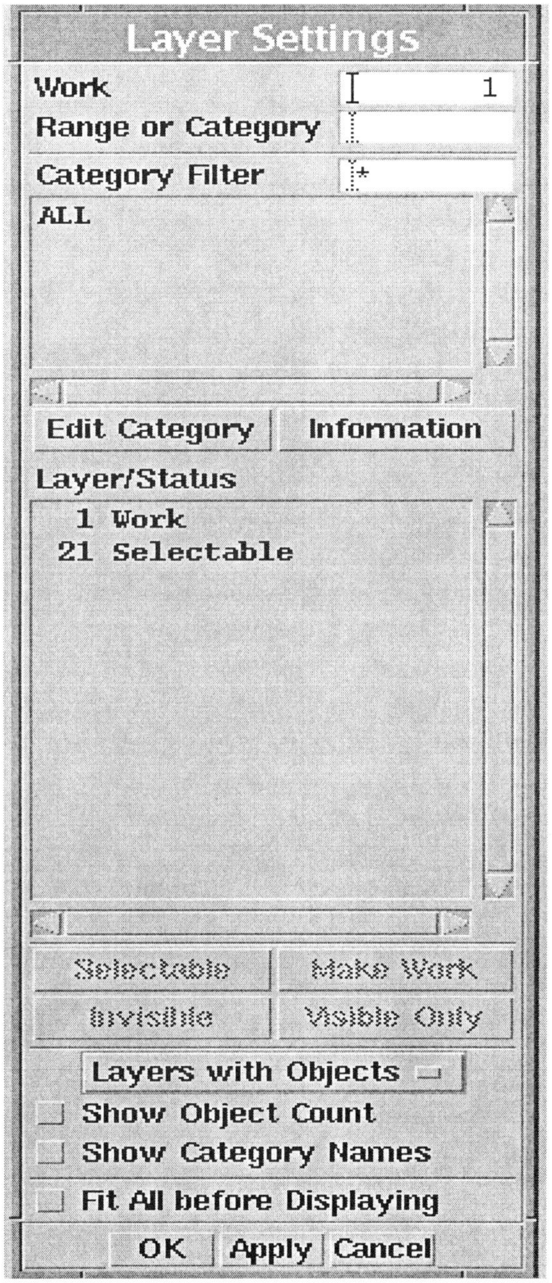

If you construct something on a layer and you want to move it to a different layer use Format|Move to Layer. Key in the destination layer and it will now the information will reside on the new layer.

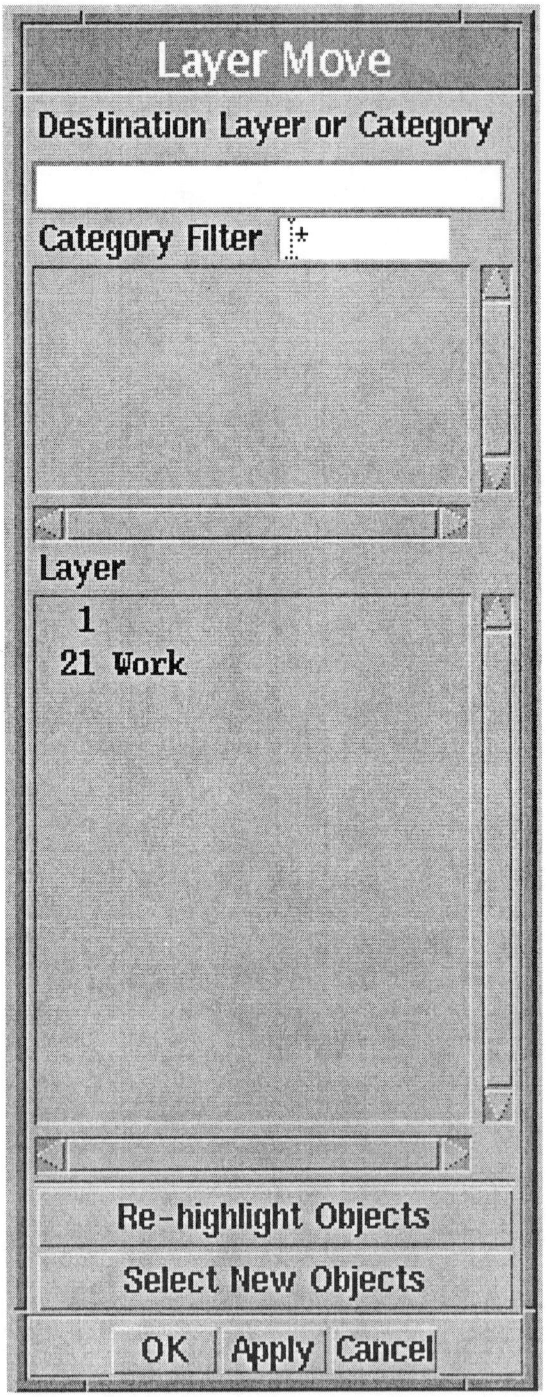

1.2.4 Sketcher

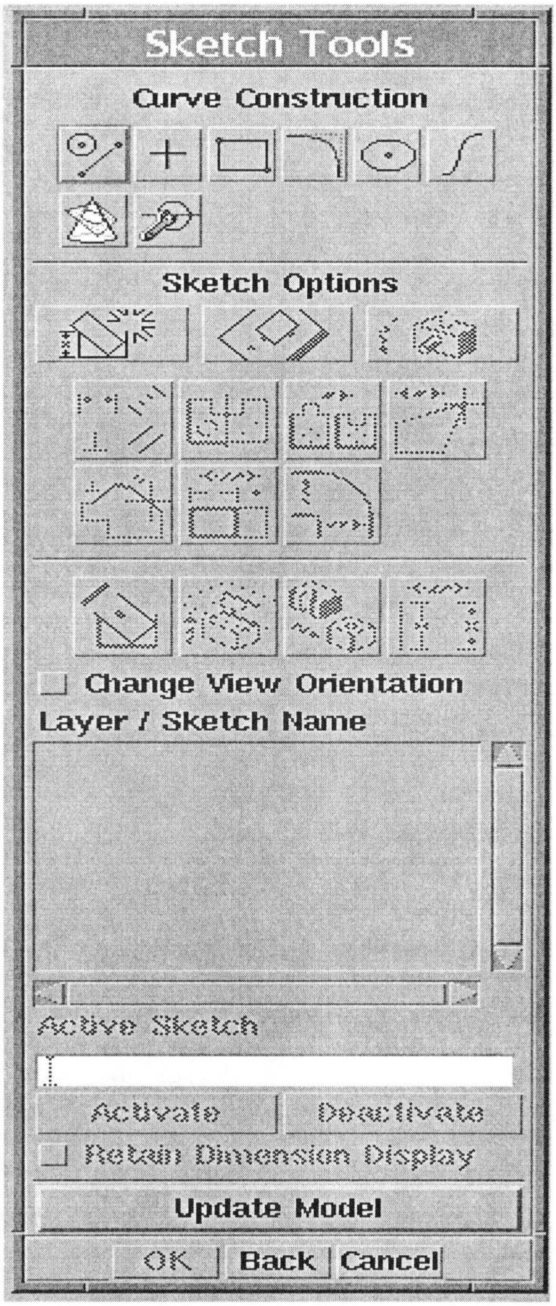

We next need to produce one of the arms of the base. The best tool for this is Sketcher. Sketcher is accessed by Insert | Sketcher. Sketcher allows us to treat the computer like a piece of paper that we can draw on. This is considered constraint based modeling.

With constraint based modeling we put, or assign, rules to our geometry. We can impose dimensional or geometric constraints. An important aspect of sketcher is our ability to not need exact dimensions to model. Because the constraints are associative and parametric to the model any changes made to the sketch will propagate to the model.

Sketcher always draws on the XC-YC plane, never in the ZC direction. We would like to draw on the bottom face of the cylinder that we just created. If you create a sketch on a plane, three helpful pieces of geometry are created a fixed datum plane and two fixed datum axes. Whenever possible it is preferable to use this geometry to make positioning easier.

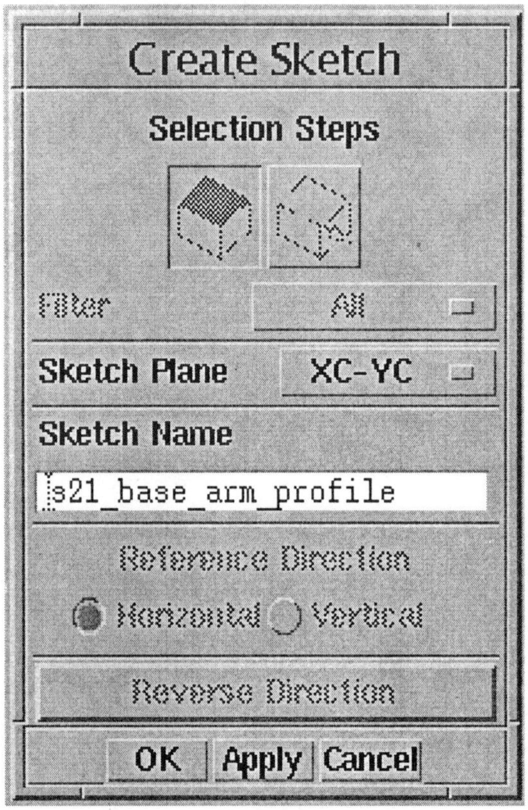

Select the Create icon, in the Sketch Options area. Because there is a piece of geometry already in the part file, you are offered a face to sketch on, if there was no geometry the xc-yc plane would be the default. You are also given the opportunity to change the name of the sketch.

A suggestion would be to change the name to something like s, for sketch, 21, for the number of the layer that the sketch is on, and a descriptive name. In our case we will call the sketch s21_base_arm_profile. Choose the xc-yc plane. You must change the layer number before you create the sketch so lets do that now. Change the layer to 21 in the work layer dialog box. Select OK.

A sketch was created with the new name with the reference geometry, but it is in an awkward view orientation. Replace the view with TOP. In the top left corner of the dialog box choose Basic Curves. With Basic curves you can draw lines, arcs, circles, and apply fillets, (corners with tangents).

Lets start with a vertical line on the right side of the vertical datum axis, the line with the arrow head on it. Because the sketch preferences default snap angle is set to 3 deg., the line will snap to vertical if it is within 3 deg of vertical, the same with horizontal. This value can be changed in Preferences|Sketch.

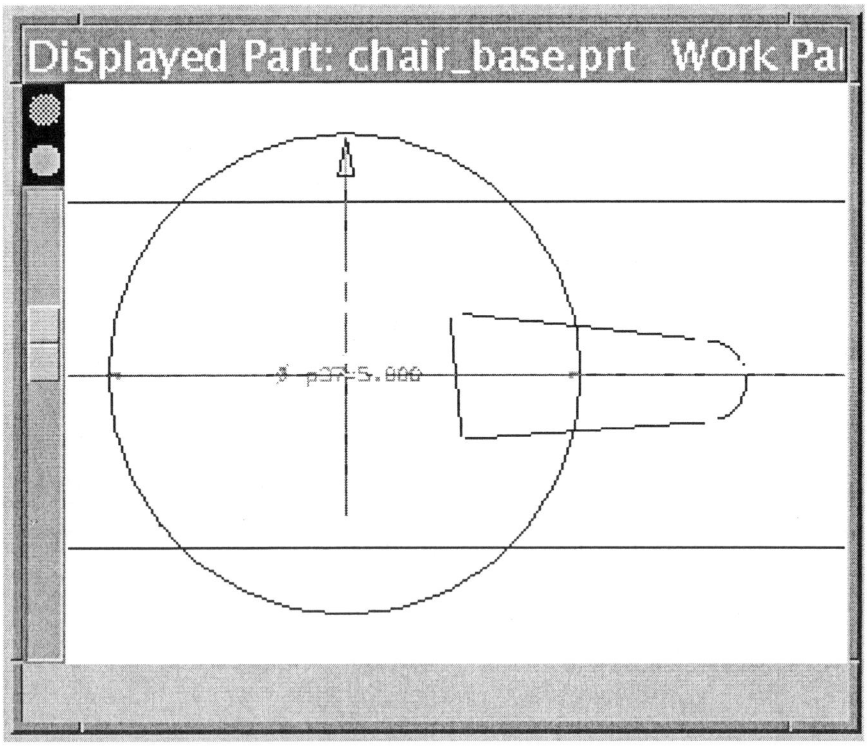

Continue on with a line to the right an arc and another line connecting with the start of the first line. Cancel the Basic Curves dialog box. We will now constrain the sketch by applying geometric and dimensional rules.

Find the Constraints icon, (the one with dimension leaders on it), and choose Geometric.The order is not that important, this is a suggestion. As many geometric constraints as possible should be applied first to help get the final shape. We need to put the first vertical line at the same place as the vertical axis.

With your cursor choose the axis, it highlights white to let you know that it has been chosen. A list of possible constraints is listed in the dialog box. we also want to choose the first vertical line. A new list is generated. This list is what the system can apply to both lines at the same time. Choose Collinear. This will move the line on top of each other. Choose Apply.

Next we need to put the arc center on the horizontal datum axis. Choose the datum axis, choose the arc center, choose point on curve, choose apply. The center of the arc moves to the axis, and can move anywhere on the axis. The ends of the arc now need to be made tangent to the angled lines. Choose the arc, choose a line, choose tangent, choose apply. Do the same to the other side.

That is all the geometric constraints we need for now. Lets switch to dimensional constraints to put some measurements on our sketch. The default is for the inferred selector. This selector lets you specify which type of dimension you apply by the position of the dimension. Lets begin by giving the vertical line a length. If the cursor can see more that one object, (it can because the line and the datum axis are in the same place), a quick pick dialog appears. Scroll to the number that represents the line and select it. A dimension appears and you must then place the dimension. Move it to the left away from the sketch.

The length of the line should be 1.5". Enter it in the current expression area and press enter to register the value. It changes to 1.5". We also need to make the distance of the line endpoints the same distance from the horizontal datum axis. The expression that was created is p2=1.5.

Remember that the cylinder was created with a diameter, p0, and a height, p1. If we put a dimension from the horizontal datum to an end point of the first line of p3=p2/2, the line would always be centered. Lets choose the horizontal datum and the lower end point. Enter a value of p2/2, press enter and the model will update. The side of the arm needs to be set to 2 deg., from horizontal, choose the axis and the line and assign the angle value.

Next the distance from the arc center to the center of the cylinder need to be set. Choose the vertical datum axis and the arc center. Change that value to 13".You will need to fit the view again. That is all that is needed to sketch the profile of the arm.

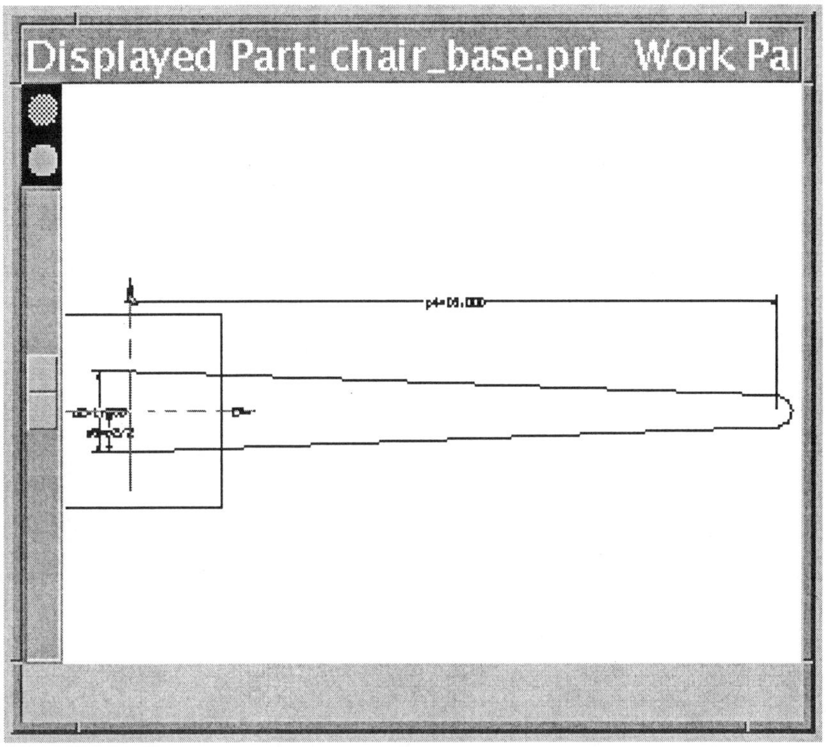

This is what the sketch should look like. Lets switch back to the TFR-TRI view and sweep the sketch.

1.2.5 Extrude Body

Sweeping has three choices, Extruded Body, Revolved Body and Swept Along a Guide. We will use Extruded Body for this part. Choose Insert|Form Feature|Extrude. The cue line wants you to select the section string. The section string is the geometry that you will sweep. In this case it's the sketch that you just produced. Choose the cyan section string of the sketch. It highlights white.

The extrusion method is the next question that has to be answered. Direction and distance is what we want. A positive direction vector is displayed. In this case we want to start a distance from the bottom and continue up a distance. Therefore the start distance is set to .5" and the end distance is set to 1.5" to make a 1" thick part. If the direction vector arrow is pointing up, accept the default direction. Key in .5" for the start distance and 1.5" for the end distance.

We won't need an offset for this part so leave them at 0. We will be creating a solid so be sure to switch back to layer 1 before we select OK. select OK, we are asked which Boolean operation we want. We want to create a single solid part so choose unite and OK. A solid body is created.

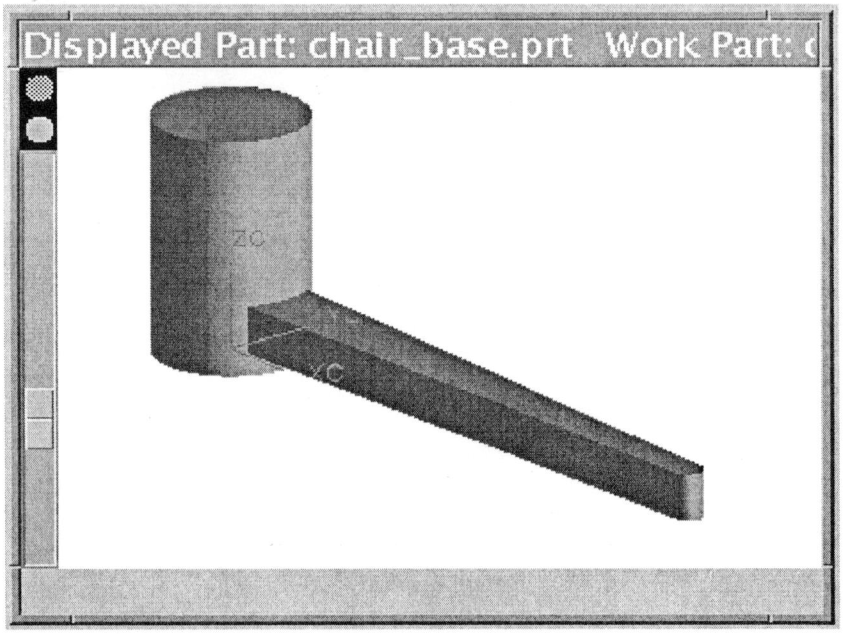

If you select Format|Layer Settings and make layer 21 invisible it will hide the sketch geometry. This is the shaded view of what we have so far.

In a moment we will create some copies of this arm but first we will need to add some features to this arm. On the outside end of this arm we need to add a feature called a Boss. Pads and Bosses are features that extend away from a part adding material.

1.2.6 Form Features

Whereas Holes and Pockets cut into the part removing material. Rotate the part around to see the bottom of the arm. Choose Insert|Form Feature|Boss. The cue line wants you to select the planar placement face. The position of the cursor dictates where the center of the feature will be placed. Choose near the end of the arm, on the bottom face.

The quick pick dialog will be available because the cursor can see more than one face or edge. Scroll through until you select the face you want. Enter the boss parameters. We will use a diameter of .5" and a height of .375".

We then need to position the boss. Since we want to position the boss at the center of the arc we can use the point onto point method. Choose the outside arc of the arm as the target edge, the target edge is the edge that you are measuring too, and choose arc center. A boss is needed on the top of the arm to align with the boss that we just made. The create boss dialog is still active.

Choose the top face of the arm as the planar placement face. This time make the parameters .5" diameter and a height of .25", and position it just like you did the bottom boss. Next we need to put a hole in the end of the arm that the casters will fit in. We will make the casters later. Choose Insert|Form Feature|Hole. The choose hole type dialog comes up and we will make a simple hole.

Again choose the bottom of the bottom boss for the planar placement face. The diameter is .375" and the length is 1.5". Position the hole with the same method that you used for the bosses. Since this is the only hole that we will need for now, cancel the hole dialog. You can shade the view at any time to view the part. Put the view back to TFR-TRI, the view will automatically fit to the screen.

What we need to do now is to create copies of the arms to make a total of 5. Unigraphics calls this a circular array. Choose Insert|Feature Operation|Instance. Choose Circular Array. The Select features dialog appears choose the extruded feature, both bosses and the simple hole, select OK. The features highlight to verify that this is what you want to array.

Next we have to enter the Circular array parameters. We want 5 total arms, equally spaced. Enter 5 for number and 360/5 for the angle. We must then select a rotation axis using point and direction. We will be using the WCS which is in the center of the cylinder. Choose OK the vector constructor appears again. Choose the positive zc direction.

You must then specify the point on the zc axis that you will use, which will be 0,0,0. A preview will appear, if this is what you want choose OK, (if not choose no). Cancel the instance dialog.

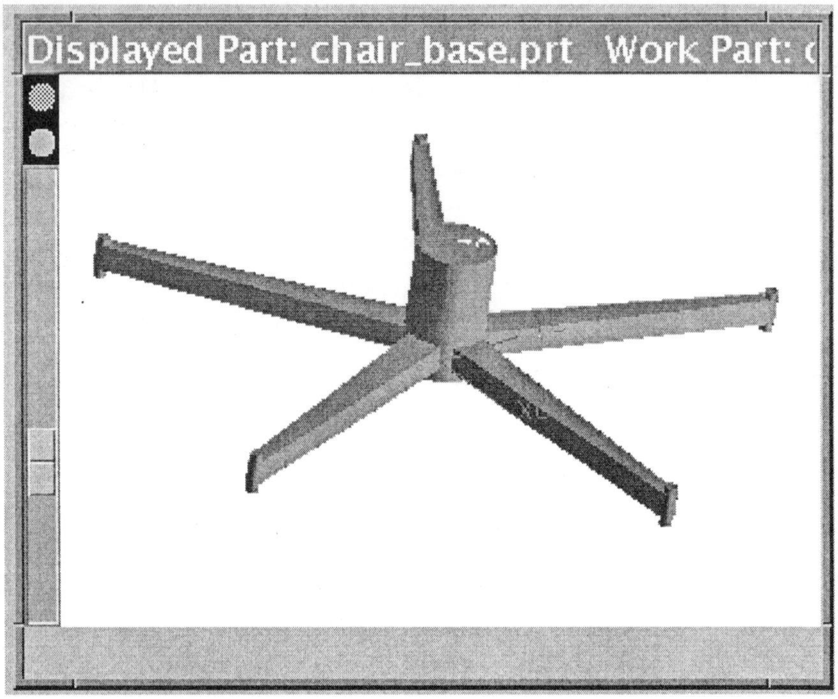

The base now has 5 arms that are attached to the original cylinder. We will now clean up the part by adding some edge blends.

1.2.7 Edge Blends

Remember that the instance that you just made is associative, meaning that if you make a change to one of the arms the other arms will also change.

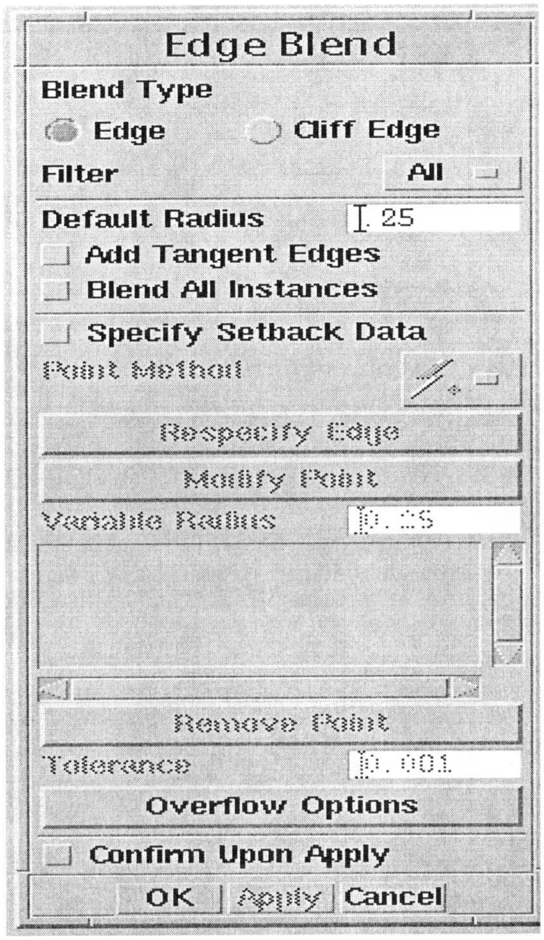

Choose Insert|Feature Operation|Edge Blend. The dialog box appears and asks us to select the edges that we want blended and specify parameters of the blends. Edge blends are like round overs of the edges. Change the default radius to .375" and make sure that you choose the Blend All Instances button. Select Apply because we will make more edge blends.

Change the default radius to .125" select Blend All Instances and Add Tangent Edges. Select a top and a bottom edge of one of the arms. Choose Apply.

We will now blend the top edges of the top boss on the arm, and the top edge of the original cylinder. Leave the radius at .125" and select the top edge. remember to select blend all instances.

Now blend the bottom edge of the original cylinder to .25". This time choose OK because we are finished with blending for now.

All we need to complete this component is a hole thru the cylinder to accept the seat post the will be made later in the book.

1.2.8 Form Features

This time make a simple hole that goes all the way thru the entire cylinder. Choose Insert|Form Feature|Hole. This time we want a simple thru hole. Select the top face as the planar placement face and the bottom face as the thru face. Enter 1.5" for the diameter. Position the hole as you did with the bosses.

If you have not already, make sure you save your parts, and save them often. This component should be saved as chair_base. You should now have one component saved named chair_base.

1.3 The Model Navigator, or MNT

1.3.1 The MNT

We also have a very easy way to manage most of the information that we have about a component, or model. The model navigator is accessed with View|Model Navigator. As you can see, all information about the model is listed in the order of creation. When you select a node of the MNT, (a feature), with your left mouse button, the feature will highlight. Likewise you can select a feature in the graphics area and it will highlight in the MNT.

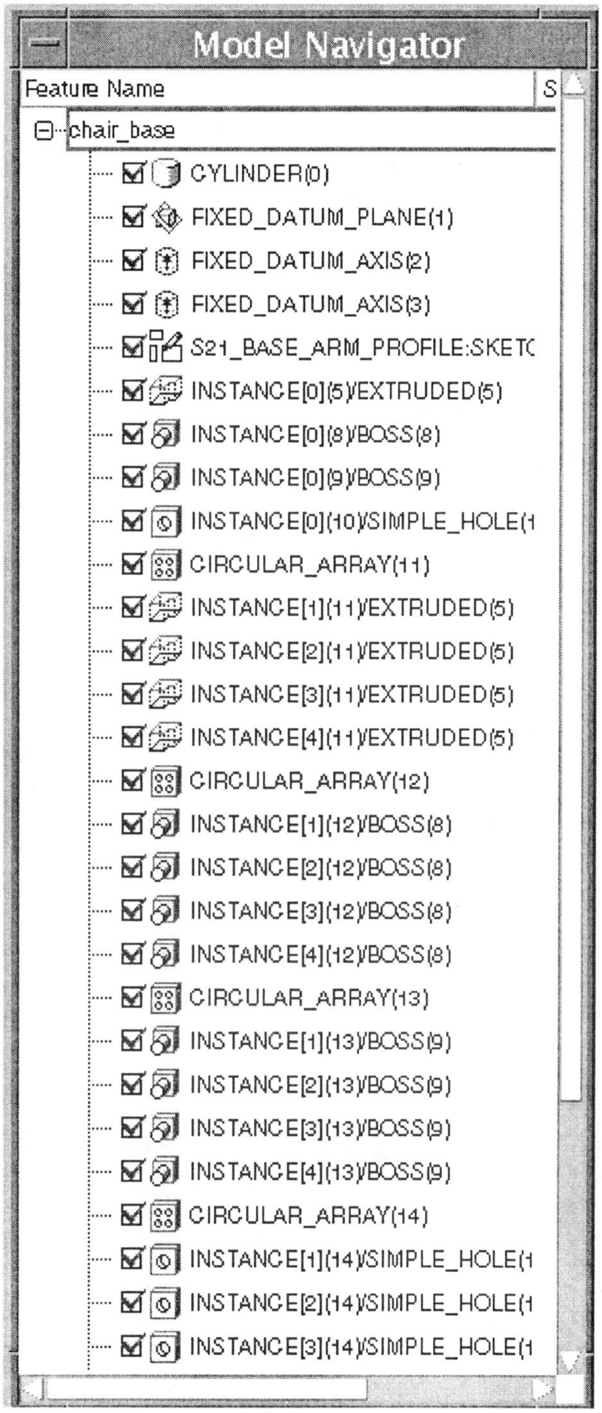

With your right mouse button, you have the avbility to access many different function that can be applied to the model feature. Choose the first feature, the cylinder, and note the available otiopns. Choose, with the right mouse button, cylinder, edit parameters, hightlight feature dialog in the edit parameters dialog, the edit parameters dialog changes to include the parameters of the cylinder. Lets change the diameter of the cylinder to 6", select OK twice to accept the changes. Lets change the diameter back to 3".

The feature you select determines the options that are available with the MNT. A helpful hint is to leave the MNT active but put it somewhere on your desktop where it is not obstruction your view of the model.

1.4 The Chair Base Wheels

1.4.1The Center Stem

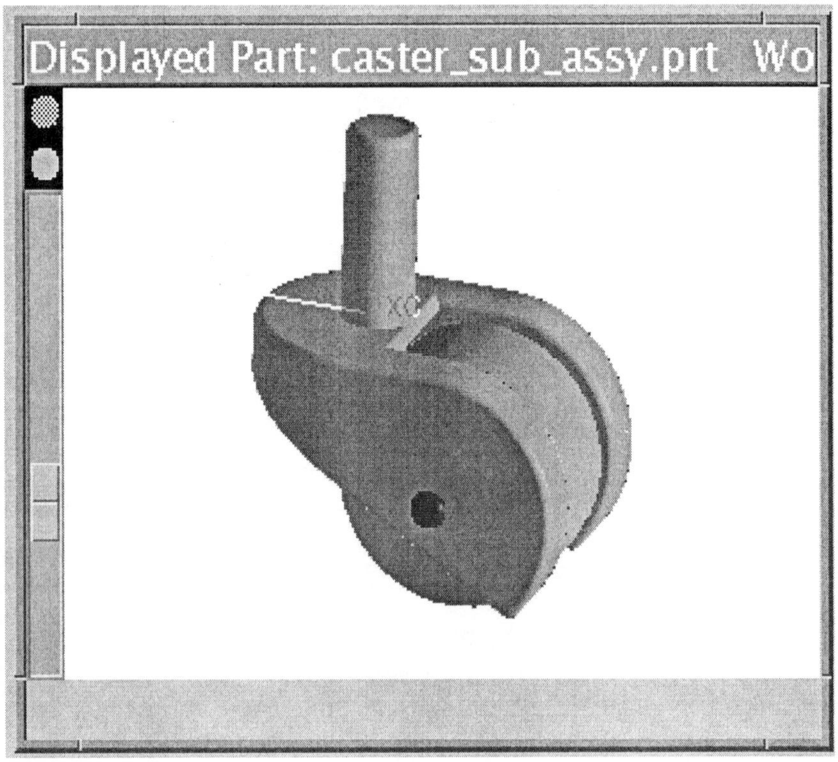

Open seed_in.prt, and enter the modeling application. The caster that we will model will be made up of a set of components that we will learn how to assemble later in the book. Lets start with the center stem that will go into the hole at the bottom of the base arm and hold the wheel.

We will start with a sketch of the profile of the part that hold the wheel. Lets start sketcher and be sure to put it on layer 21. For the purposes of this book lets use layers 21 - 40 as the standard that we will be using. Reproduce the drawing that is illustrated here. Fully constrain the sketch with the geometric and dimensional constraints necessary.

This is a simple shape that uses four lines and two arcs. The arcs are two arcs that are connected end to end with the left hand end point constrained to the angled line with point onto line, and three dimensions, (your instructor will walk you through this drawing).

Remember to draw the shape freehand and then move it into position with constraints. Deactivate the sketch to look at in the Trimetric view. Lets extrude the section string to create a solid that will become the thickness of the center

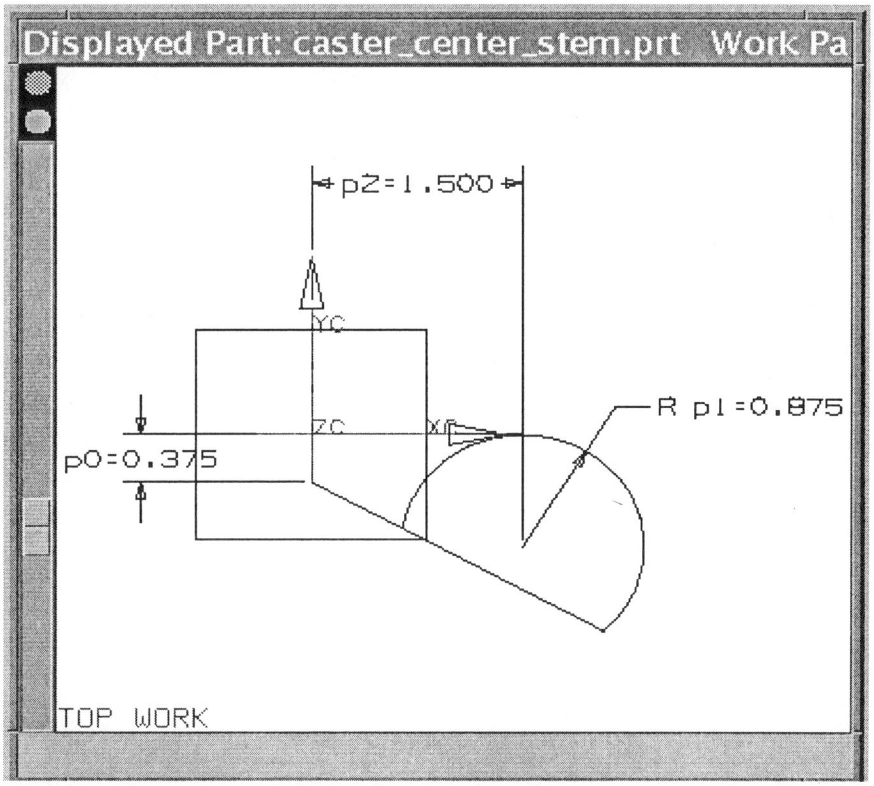

flange. We will need this part to be .875" thick, but the datum plane from the sketch need to be centered on the part. That means that our start distance is -.875/2 and our end distance is .875/2. Since we really have two sketches, if we selected the section string all lines and arcs would be selected.

We only want the left 4 curves selected. Lets choose curves in the select section string dialog box and select the four curves that we want, select OK, twice. If the direction vector arrow is pointing up accept this direction and enter the thickness parameters, select OK.

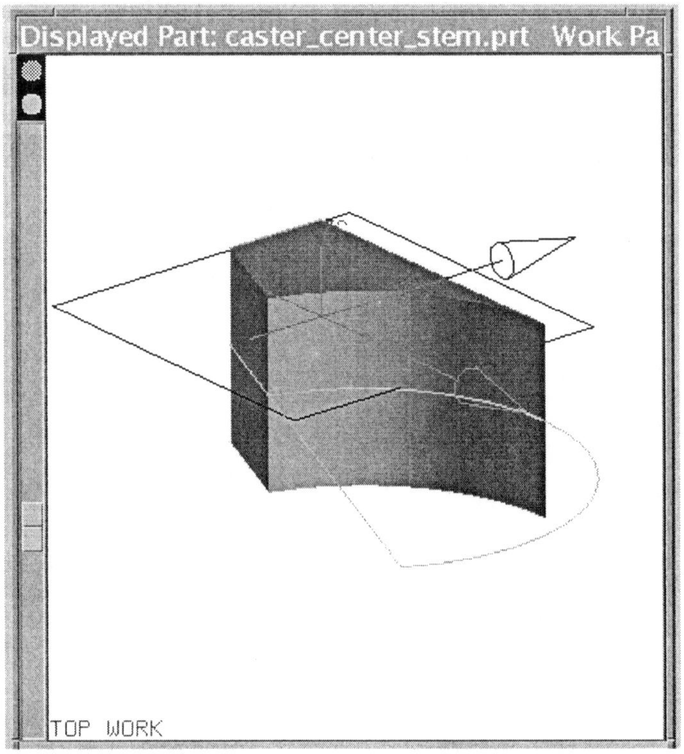

This is the shape that will be reproduced. The outside shapes are the next to add. using the outside sketch curves we will not extrude the shape. The select section string dialog box should still be active. Select all the outside curves, select OK twice. select direction and distance. If the arrow is pointing up accept it and enter the parameters.

We want to start the extrusion where the last one ended and we want a thickness of .375". Select OK. Because these parts need to be joined together, choose unite. Create the other side of this part and adjust the parameters accordingly. Cancel the Extruded body dialog box.

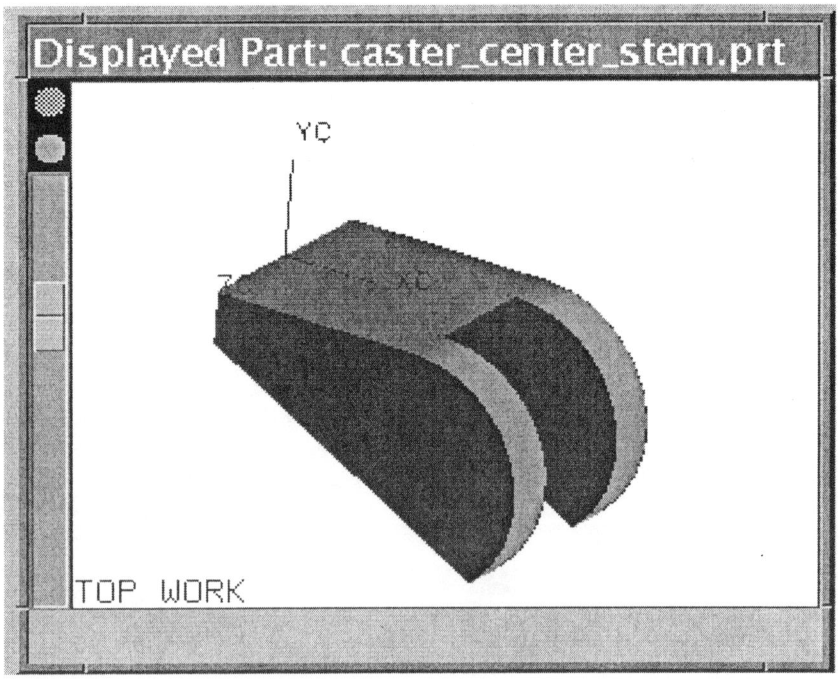

Lets now round over the back edge with an edge blend of .8175". Choose the two vertical edges and select apply.

Blend all remaining edges to .0625"

Now we need to construct the stem that will go into the hole in the base.Rremember that the hole was .5" in diameter and 1.5" long. For our purposes we will assume that clearance is not a problem. We need to add a feature that adds material to the part and is circular in shape. Lets choose a boss. The planar placement face is the top of the part. remember that where you pick the face with the cursor is where the center of the feature will go.

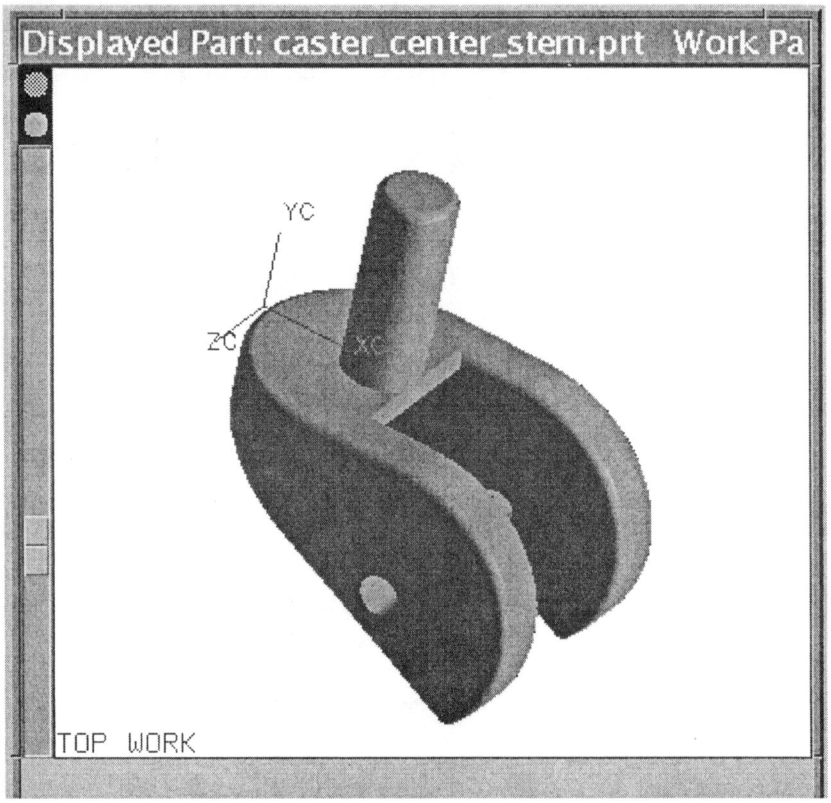

Now we need a hole for the axle to hold the wheel in place. Let make a .25" diameter thru hole from the outside face to the opposite outside face. Position the center of the hole at the same spot as the arc center of the large arc.

That is it for this part. next we will make the wheel itself.

1.4.2 The Wheel

The wheel is simply a cylinder with the outside edges blended and a hole through the middle.

Open seed_in and enter the modeling application. Remember how to make the cylinder from the very first thing you did? Insert|Form Feature|Cylinder. This time lets orient the cylinder in the positive YC direction. Blend the outside edges to .0625". Put a thru hole in the center of the wheel of .25". Position the hole in the center of the wheel. Close the create hole dialog.

Hopefully you have noticed that the default color of the solids that you have created is green. Later, when we create an assembly, if everything that we assemble is green it would be hard to tell some parts apart. Lets edit the object display.

1.4.3 Editing the Object Display

Choose Edit|Object display. Select the wheel, choose OK. the edit object display dialog box appears. You can change, or edit, many attributes of how an object is displayed. We are most interested in changing the color of the object to make it more easily distinguishable from the other components. Select the color tab. The color dialog appears, choose a new color for the wheel. Lets choose an orange color. The new color will now be in the color tab. Select OK to change the color of the object. Save this part as caster_wheel.

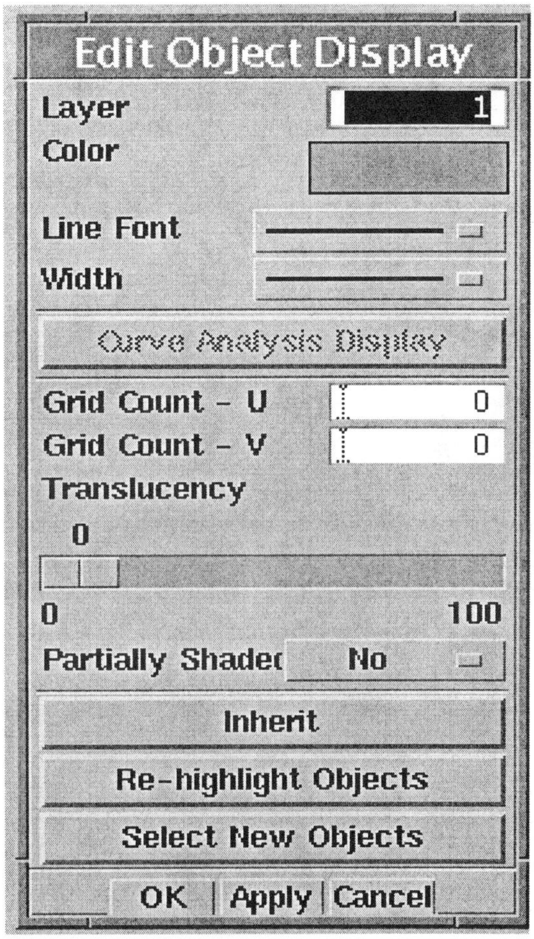

Lets go back to the part, caster_center_stem and change this to a color of your choice.

1.4.4 The Caster Axle

Open seed_in, and enter the modeling application. We are making certain assumptions about how parts are really made to simplify modeling for the beginner.

This part is simply a cylinder with the end faces blended. Lets build a cylinder that has a diameter of .25" and a length of 1.75". the blend on the ends will be .0625". lets change the color to a dark blue. Save you part as caster_axle.

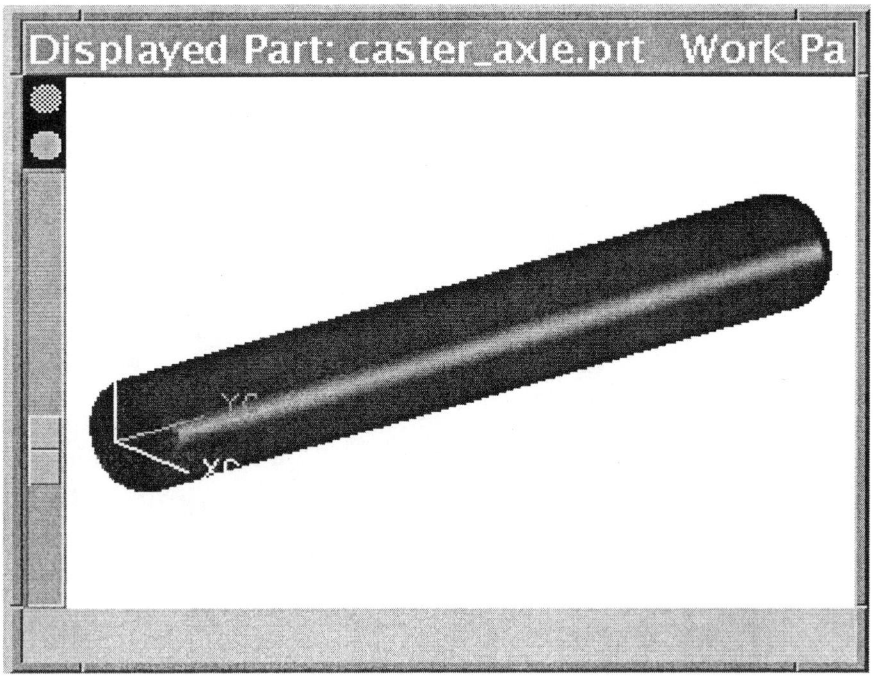

1.5 The Chair Pedestal

1.5.1 The Center Post

The center post is the part that will hold the seat bottom and also allow the seat to rotate. We need to have a cylindrical shape that goes into the chair base.

Remember the hole that we put in the base? It is 1.5" diameter and the length of the cylinder. We will start with a cylinder in the positive ZC direction that is 1.5" in diameter and 4.5" in height. Be sure to place the center of the cylinder at 0,0,0. Now place a boss on the top face of the cylinder that you just created, with a diameter of 3" and a length of 8". we will trim the top of the cylinder off later.

We now need a sketch that we can use to make the part that the seat will attach to. It has to be positioned 10" from the current position of the WCS. Choose WCS|Origin. Move the WCS in the positive ZC direction 10".

Remember that when you use sketcher, you only draw on the XC-YC plane. We must therefore rotate the WCS. The YC

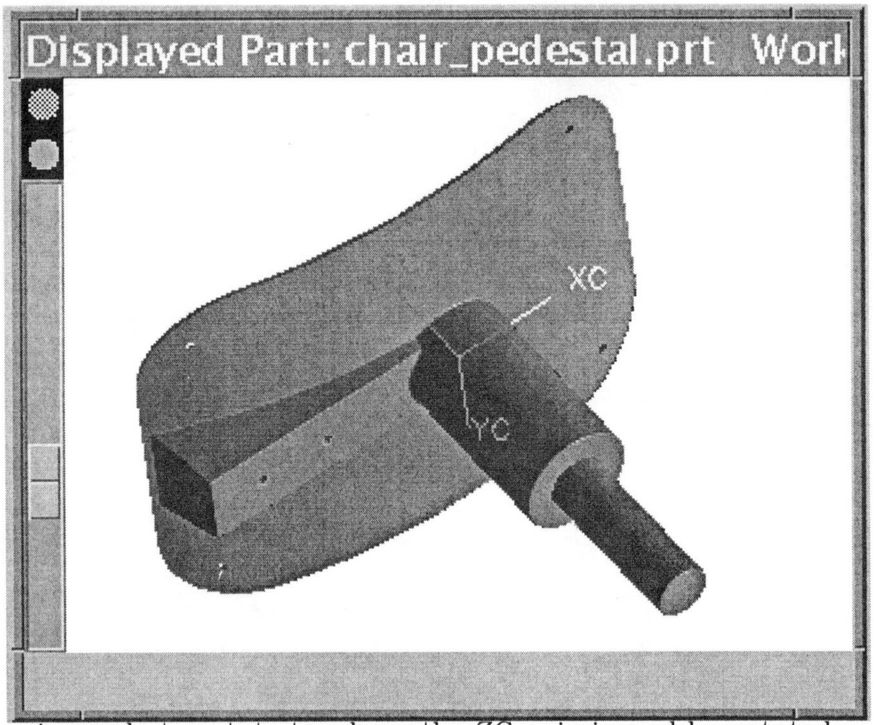

axis needs to rotate to where the ZC axis is and be rotated about the XC axis. Choose WCS|Dynamic. Grab the ball that will rotate the WCS in the direction that we need and reposition it.

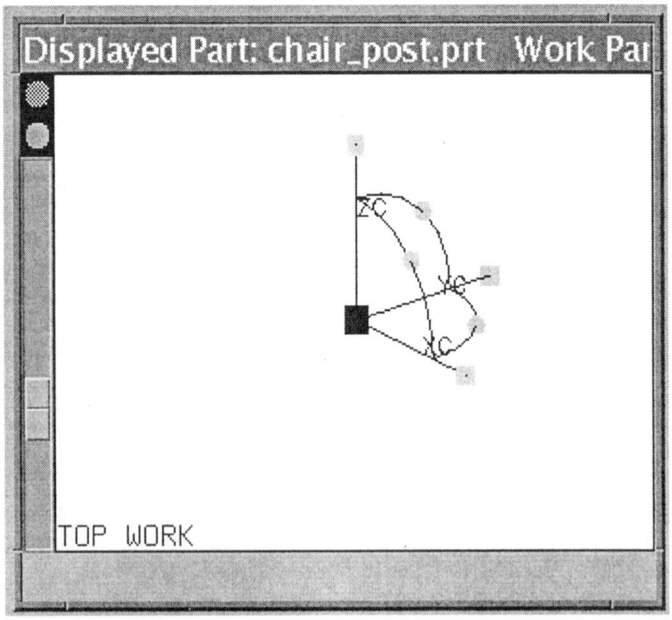

Choose Insert|Sketch. Create a sketch on layer 21, name it s21_seat_mount and place it on the XC-YC plane, choose OK. If we change the view orientation the sketch plane will be easier to see and draw on. Select Change View Orientation and reactivate the sketch. Fit the view.

Since we will be drawing on the existing part and some strange things could happen we need to make the part that is on layer one visible only. Select Format|Layer Settings make layer 1 visible only, choose OK.

Draw the shape that is depicted in the picture. It consists of one arc whose arc center is constrained to the vertical datum axis. dimension the arc with a 36" radius. Make the arc tangent to the horizontal datum axis. The right end point is 8" from the vertical datum axis, and the end point of the arc are 18" apart. Choose OK to cancel the constraints dialog. Choose cancel to dismiss the sketch tools dialog box.

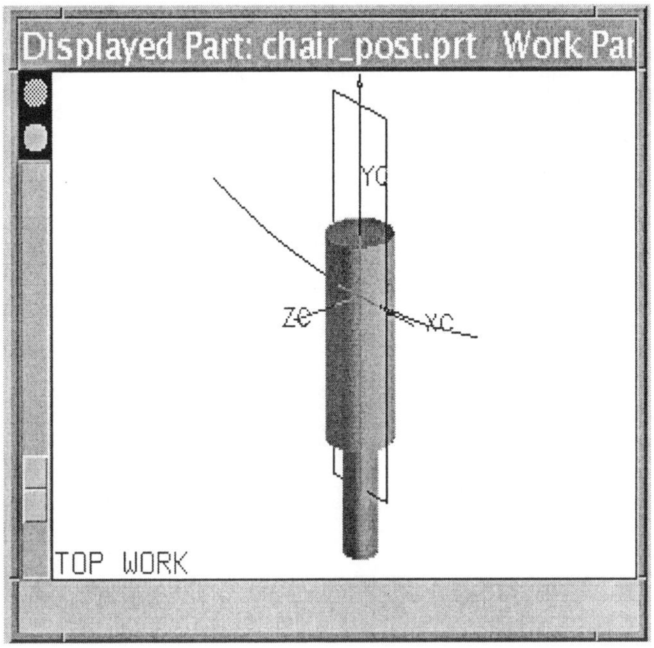

Lets extrude the section string that we just created. Here we will be using the offset feature of the extrude command to produce thickness for the part. Choose Extruded Body, select the string, select OK, select direction and distance, select OK, accept the default direction if is pointing to the left, select OK. The width of this feature needs to be 12" centered on the post. The start distance would then be -6" and the end distance would be 6". The thickness is produced by adding a measurement in the offset. The direction of the dashed arrow dictates the positive direction vector of the offset. In the illustration the arrow is pointing down. This is the direction that we need to go. We need to start at the section and offset it down a distance of .125". select OK. The create Boolean operation needs to be selected because we need to do an operation later that need this to be a separate part for now. Select OK, the shape is formed. Cancel the extrude body dialog box.

Make layer 1 the work layer, this automatically makes layer 21 selectable.

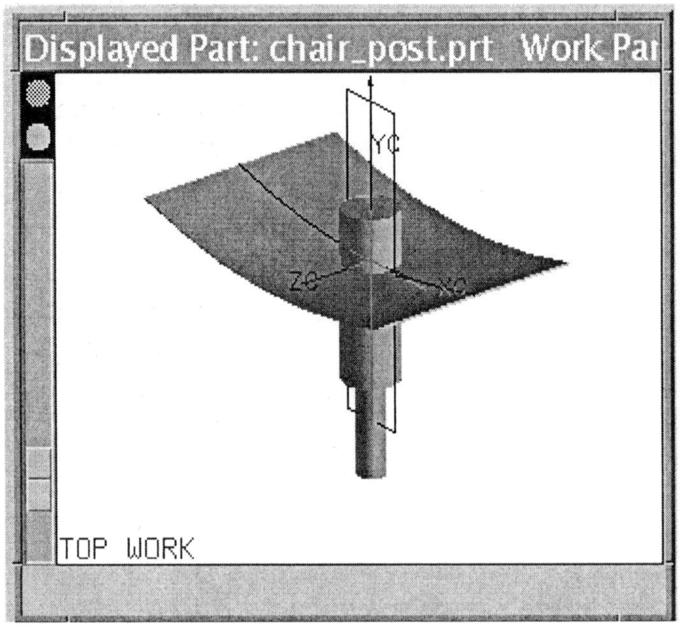

The top of the post needs to be trimmed off. Using the top face of the swept feature, we can trim off the post.

Select Insert|Feature Operation|Trim. The trim body dialog box appears, the cue line asks you to select the target body. remember the target body is the body that you will be trimming. Select the boss. Since this is the only body that we will be trimming now select OK to confirm that this was the last body. The trim body dialog box asks you to select the face or datum plane to trim with. Choose the face of the swept feature.

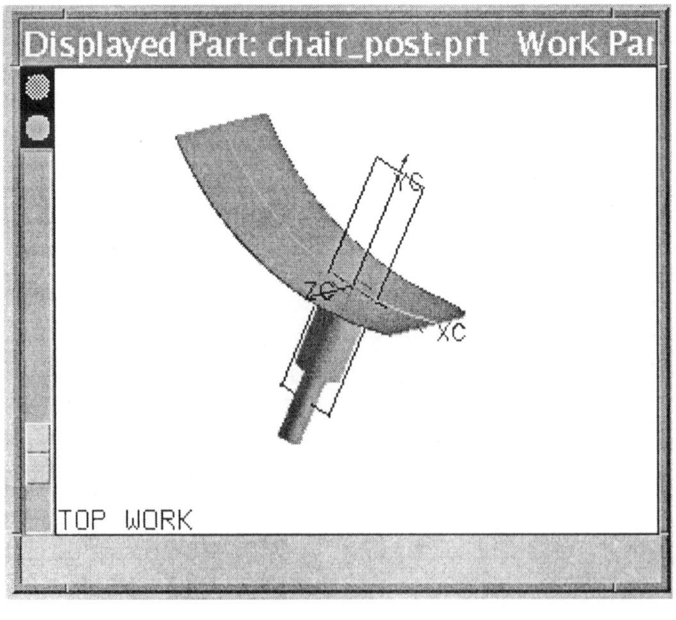

The direction vector arrow informs you of what part of the body will be removed. Select OK. If the arrow is pointing, accept this direction and choose

OK. Cancel the trim body dialog box.

1.5.2 Editing the Sketch

The measurements need to be altered to make for a better fit to the seat. Since the sketch is what is controlling the shape of the mounting pad, lets re-enter the sketch tool dialog box. Reactivate the sketch and change some parameters. Select constraints, change the value of the radius of the arc to 20", and the distance of the right end point to the datum axis to 6", and the distance from the two end points to be 14", select OK. Update the model. Cancel the sketch tool dialog box.

The corners are too sharp and need to be rounded. Edge blend will work nicely for this. Blend all corners with a 3" radius.

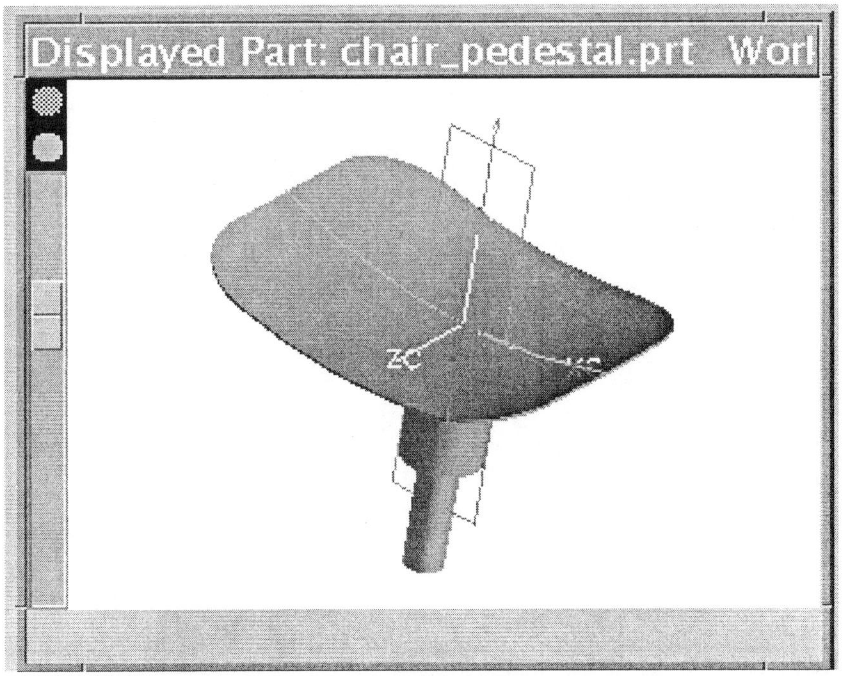

Some hole need to be put in the plate for mounting purposes.

1.5.3 Features on Non Planar Placement faces

Up until now, when we needed to place a feature on the part, we needed to select a planar placement face. The position of the mounting holes doesn't give us a planar placement face to attach a hole to.

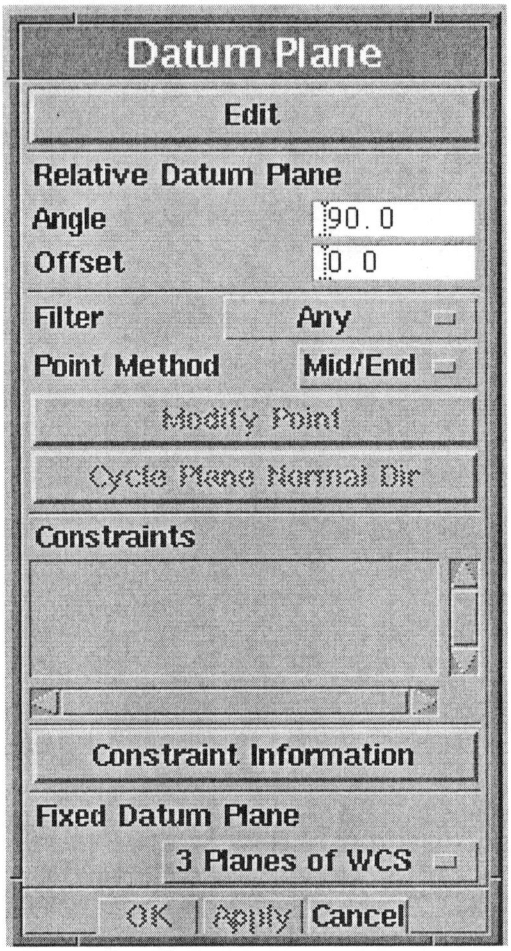

When we create a new sketch we are given three pieces of geometry that we can use to help position features and so forth. This geometry is called reference geometry. This reference geometry is also known as datum planes and datum axes. If we look at Insert|Form Feature|Datum Plane, we get the following dialog box.

If you chose apply now you would create 3 fixed datum planes relative to the WCS. If the part moved to another position the datums would stay with the WCS. What we want to make is datums that are relative to the component. That way if the part moved the datums would move with the part, or if a datum was made relative to a face or an edge and that face or edge changed, the datum would also change.

1.5.4 Reference Geometry: Datum Planes and Datum Axes

A datum plane or axis is the full size of the modeling envelope, which is 1 cubic kilometer. A border is produced so you have something to pick with the cursor. With a planar placement face where you picked on that face determined where the center of the feature was placed. With datums the feature will always go to the center of the datum.

When you make the first datum, it is always placed on the YC-ZC plane. Lets change the WCS back to it's original position. Choose WCS|Dynamics. Replace the YC axis with the ZC.

Choose Insert|Form Feature|Datum Plane, and select one of the cylindrical shapes, choose apply. a relative datum plane is created on the YC-ZC plane. Notice that you can now move the WCS and the datum plane you just created will stay.

Lets say that we now need a datum that is at a right angle to the first datum that we made. If we are making a relative datum plane it must be relative to something. If we want it relative to not only the cylinder, but also the datum, (90 deg. to it), we must choose both then apply.

When you choose both then apply, the cue line asks you to pick the required plane normal/axis vector. Five arrows are created. One in a dashed arrow and four solid arrows. The solid arrows if chosen will produce a tangent plane in the direction of the arrow. The dashed arrow will allow you to rotate the new datum relative to the first plane. Lets choose the dashed arrow and key ion 90 for the degrees. Choose apply and a new datum is created that is 90 degrees to the first one. If you choose only one datum plane, (or a planer placement face), you can then create an offset datum the direction vector arrow that is shown is the direction that a positive value will go. A negative value will go in the other direction.

For our part we need to create four datums that are offset inward 1.5" from each outside face, (the .125" side), one on each side face of the plate. Just as we used layers 21 - 40 for sketches, lets use layers 61 - 80 for reference geometry. Make layer 61 the work layer. Choose Insert|Form Feature|Datum Plane. Chose one face of the side faces. The arrow is pointing in the positive direction and the offset should be set to 0. Change this to a value of -1.5, (if the arrow is pointing out). Do that for all four faces.

These planes will cross 1.5" in from the sides in all four corners.

We now need something that is normal to the face that we want to put the hole in. Where datums, or datums and faces, cross we can construct a datum axis. Choose Insert|Form Feature|Datum Axis. Choose two crossing datums and select apply.

Do that to all four corners. We now need to make a planar placement face that is normal to the face of the plate along the datum axis that we just made. These planes need to be made relative to both the face, (either face), and the axis. Choose Insert|Form Feature|Datum Plane. select the face of the plate and one of the datum axis. Select apply.A relative datum plane will be placed above the part, (you may have to fit the view to see it). Do the same for the other side. You will only need one plane per side.

The hole type need to be thru hole so we will need another plane on the bottom of the part. Choose one of the upper planes and give it an offset of 30" and be sure to check the direction vector. Do the same for the other side. The part will look like the following.

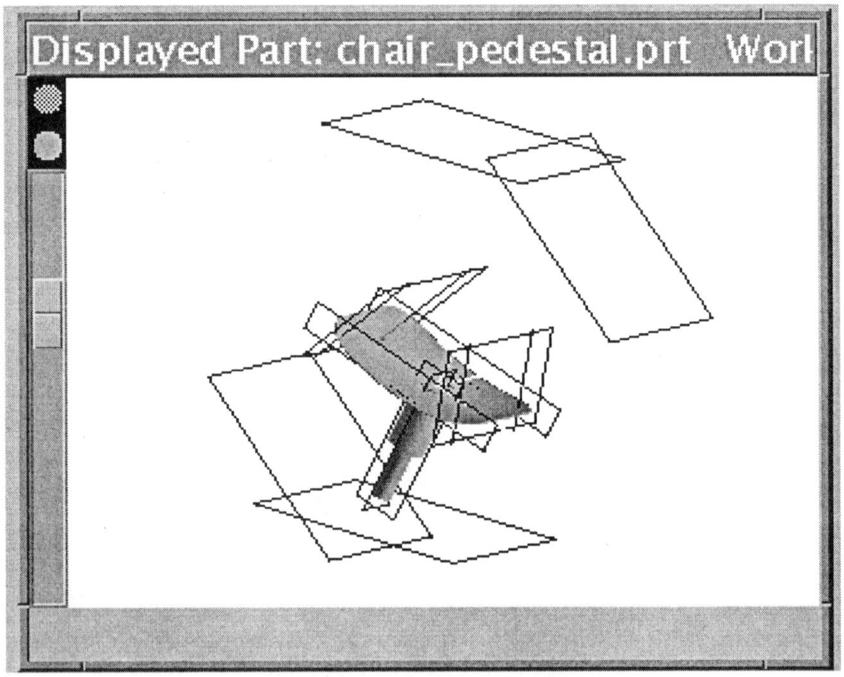

We now have enough geometry to construct the holes. This is necessary because we want to be able to change the sizes of our parts if necessary and have the holes remain associated to the design intent.

Select Insert|Form Feature|Hole. Lets use a simple thru hole. We now have planar placement faces to put the hole on. Select one of the upper datum planes as the planar placement and the opposite plane as the thru plane. Make sure that the arrow is pointing in the direction of the part. The hole need to be .25" in diameter.

We will move the tool solid that will become the hole into position. This time we need to use the center point of the hole and the datum axis for positioning. That will be point onto line. Choose point onto line and OK. Create the other three holes and position them as above.

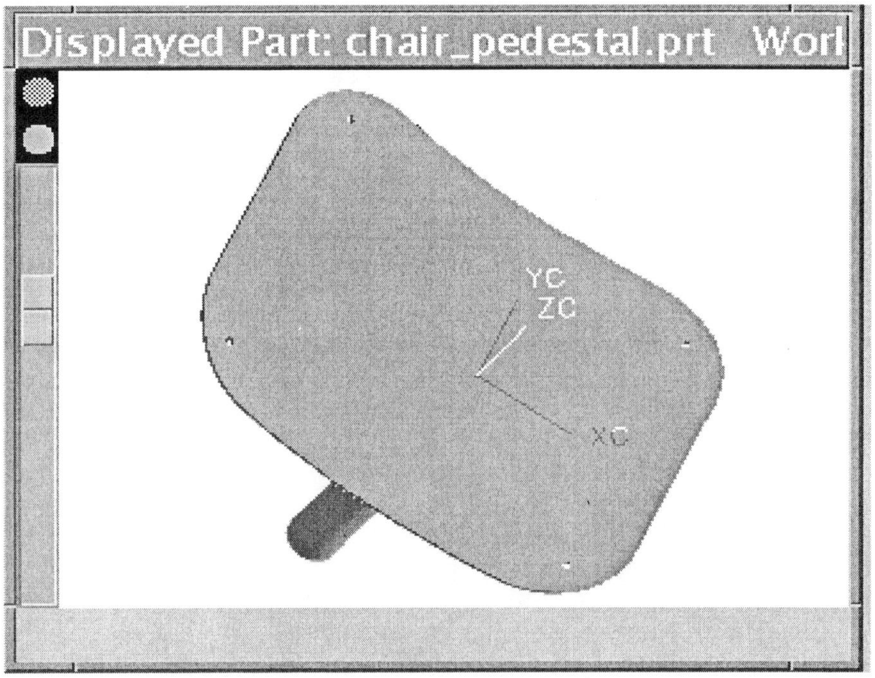

Next a pad needs to be put on the bottom of the plate at the back so we can attach the component that will hold the seat back in place. Rotate the part over to see the bottom. We will use another sketch to form the outline of the pad. In order to get the pad in the proper position we must move the WCS, remember the sketch will be created at the position of the WCS. Select WCS|Origin Choose the arc of the boss where it meets the cylinder. Make sure that the ZC Axis is pointing in the direction of the plate. Now move the WCS 5" in the positive ZC direction.

The pad will look like this.

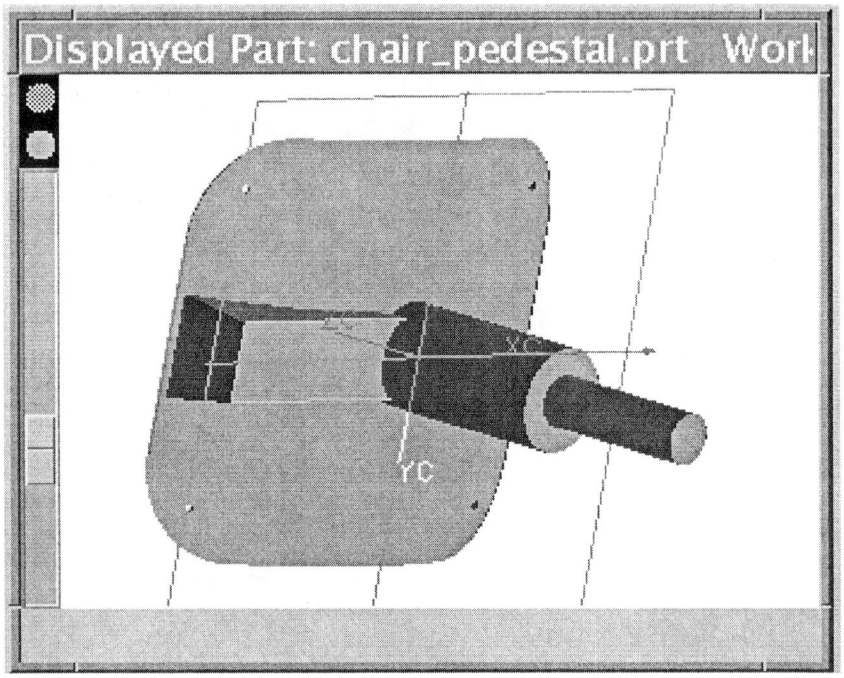

Create a sketch on the XC-YC plane. Draw a rectangle and constrain one short edge to the datum axis that goes through the center of the boss. Additionally constrain the rectangle to be in the center of the other datum axis. Add the following dimensions. The length will be 7.5", the width will be 2.25", centered on this axis.

Change the work layer back to 1 and extrude the section string that you just created. Choose Insert|Form Feature|Extrude. Select the section string, Ok the extruded body dialog box. We now want to extrude the body to the curved plate. If we used direction and distance it would stop at the first instance that the section string touched the plate. Lets choose trim to face/plane.

Check the direction vector we want to go to the plate. reverse the direction if necessary, choose OK. Next we get a Trimming face dialog. We need to extend the trim face, then choose the face of the plate. select OK. The extrusion is formed and we want to unite this part to the plate.

1.5.5 Taper

The sides of the pad that you just created need to be tapered. Tapering needs to be done to some parts so they can be removed from a mold. Choose Insert|Feature Operation|Taper. We will taper the faces of the pad. Select the three visible faces, and the default angle of 10, select OK. Next you need to select the draw direction, choose on of the small vertical edges of the pad.Make sure that the direction vector arrow points in the direction that the mold would be removed, select OK. Next the reference point is requested. The reference point is a point that will not change. Choose a point on the face of the pad, a little cross will be applied to the corner., select OK. The taper is applied.

We now need to put some threaded holes in the pad so we can bolt the back cushion support to it. remember when we put the first datum planes through the cylinder and boss? We will now use them to position the holes. Choose Insert|Form Feature|Hole. Select simple, select OK. The planar placement face will be the flat pad face.

The positioning method will be point onto line. We will use the center point of the hole and position it onto the datum plane that is parallel to the long side of the pad. We will then use point to a line at a distance, or perpendicular, for the other dimension. Use 6" away from the center for the measurement. Create another hole that is positioned the same except that it is 4" from the center this time. Cancel the hole dialog.

When this is finished make layer 1 the work layer and all other layers invisible. The only thing visible now is the part itself. We also have holes in all four corners.

1.5.6 Threads

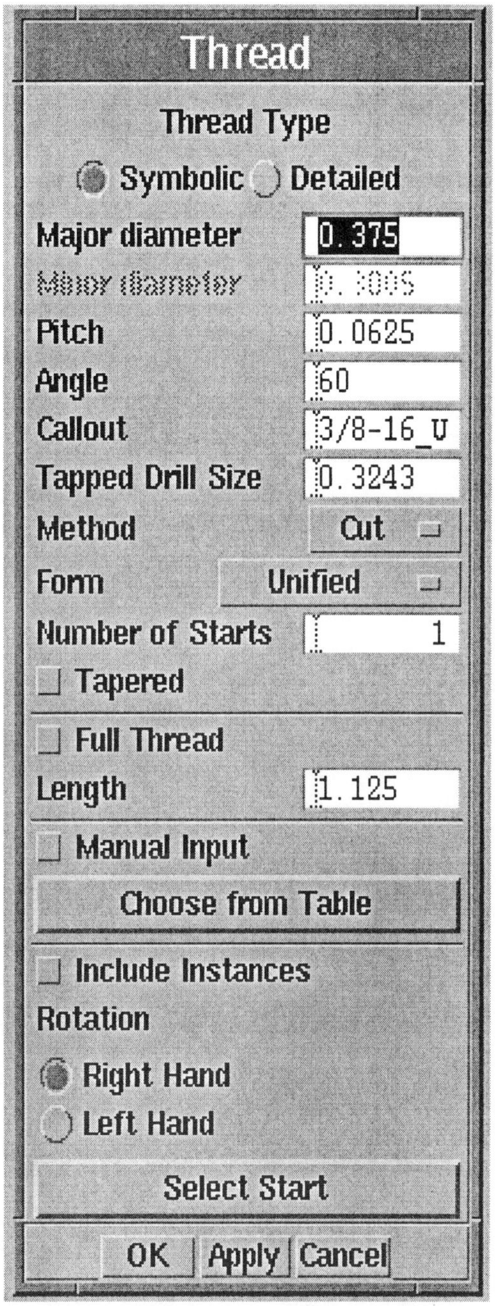

The holes that we just made need to be threaded. There are two ways to depict threads, Symbolic and Detailed. detailed threads form a true representation of the thread but little information about the thread is associative. Basically detailed threads makes a pretty picture. Detailed threads require a lot of computing power to render and display. For this purpose we will use symbolic threads when ever possible.

Choose Insert|Feature Operation|Thread. The thread dialog box appears and the cue line asks you to choose the cylindrical faces to apply the threads too. We will use symbolic threads. Choose the two holes just created. The fields fill with the information for the holes. The method box lets us choose how the threads are made, we will choose cut. The form lets us choose what type of threads are produced. We will be making 3/8-16 Unified National Course, or UNC. We can then choose from the table, look for 3/8-16 UNC and select it. All required fields are filled in with the required parameters.

These are the only threads for now so select OK to create the threads and cancel the dialog box. A symbolic thread is shown by adding a dashed line to depict either the major diameter of an internal thread or the inside of an external thread.

If a different size hole is used, the hole will resize to the required size for the specified thread needed.

Apply some.0625" blends to the edges of the part, but not to the holes or the plate itself.

Don't forget to change the color to something different than the joining parts.

Remember when we started this part and we just created the extrusion instead of uniting it? Lets now join the two parts together. Choose Insert|Feature Operation|Unite. The target solid is the part that you are adding parts too. Select the post. The tool solid is that the part or parts that you are adding. OK the class selection dialog. Cancel the unite dialog.

Choose Format|Layer Settings, and make sure that layer 1 is the work layer and all other layers are invisible. Save this part as chair_pedestal.

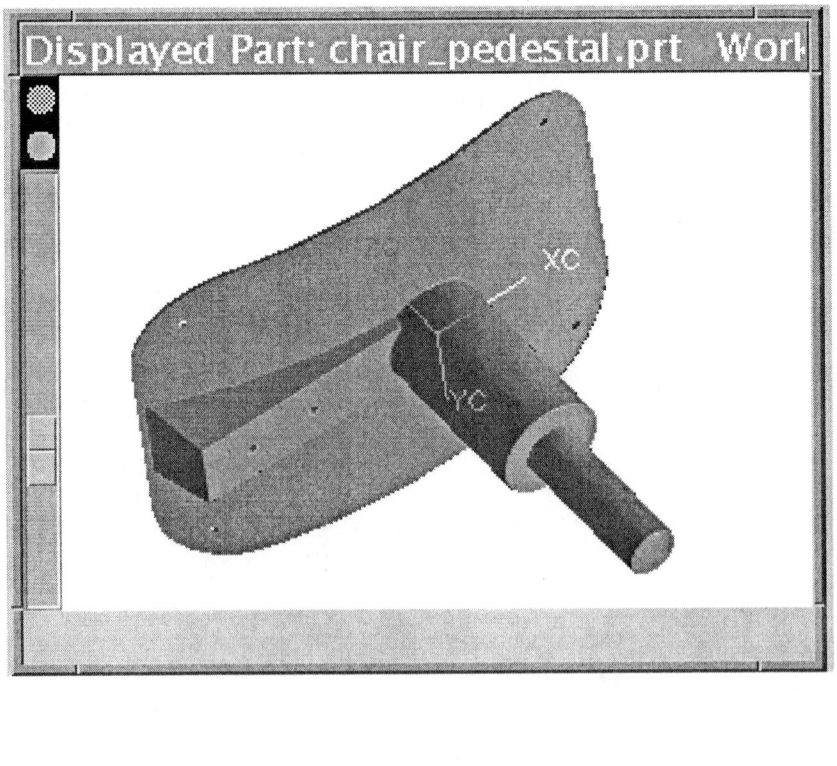

1.6 Chair Back Support

1.6.1 Support

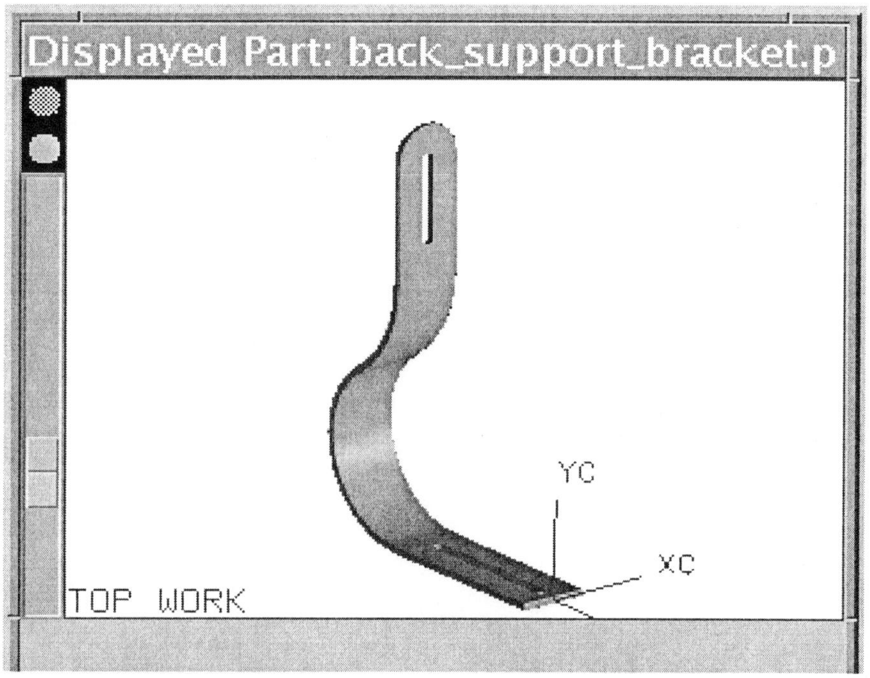

Open seed_in and start the modeling application. Because of the shape of this part a sketch is the best tool to use. Change to layer 21. Choose Insert| Sketch, Create. Change the name to s21_back_support_bracket, and choose the ZC-XC plane to draw on.

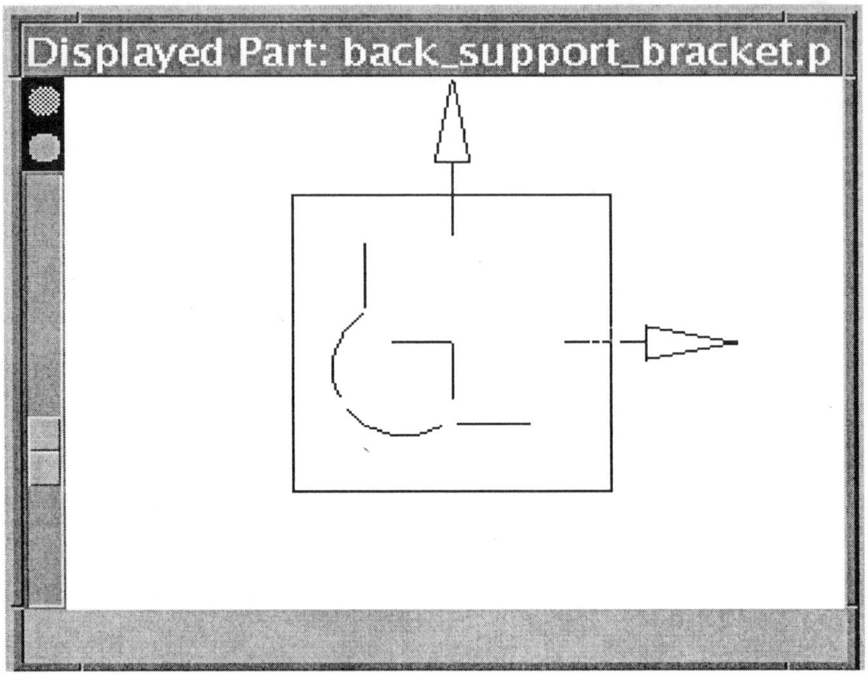

The sketch can start like this. It consists of a short vertical line an arc and a short horizontal line. We will next constrain the sketch. The bottom line and arc need to be tangent to one another. Next lets move the bottom line collinear to the horizontal datum axis and the vertical line collinear with the vertical datum axis, and the arc center on the vertical datum axis.

The length of the bottom line is to be 6", the length of the upper line is 5", and the radius of the arc is to be 3". A fillet of 4" can now be applied to the corner of the arc and the upper line. The sketch should now look like this.

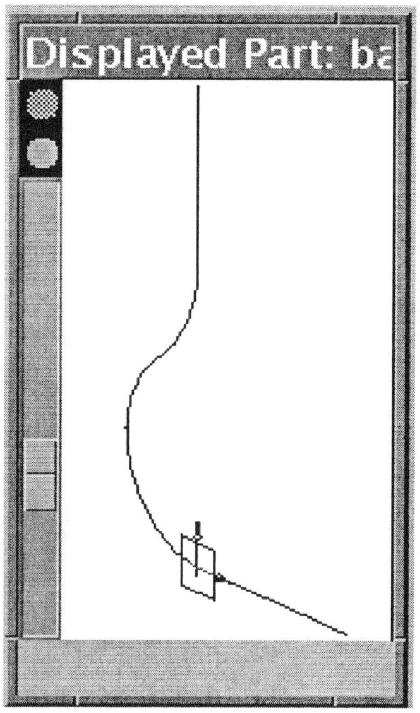

The WCS needs to be moved to the end point of the section string so we can sweep along a guide a section shape. Choose WCS|Origin, select the bottom end of the line.

1.6.2 Swept along a guide

Another sketch need to be constructed on the XC-YC plane. Change the work layer to 22, (second sketch). Choose Insert|Sketch, Create. Change name to s22_section_profile, and create it on the XC-YC plane. draw a rectangle and center it on the endpoint of the first string that you made earlier. The dimensions of the new section should be 2.25" wide and .25" high.

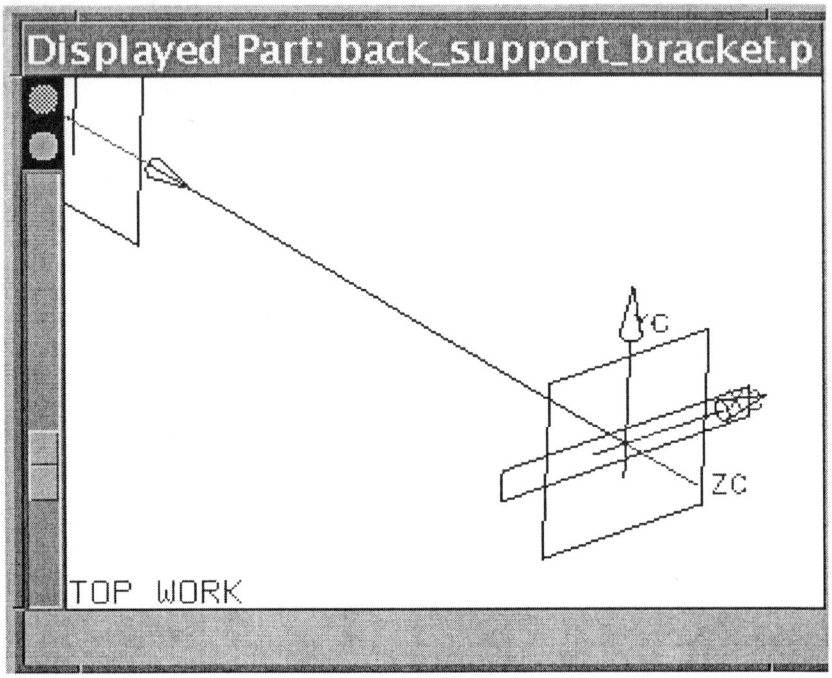

The shape that was just created will be swept along a guide. the guide we will use is the first string that we created. Choose Insert|Form Feature|Sweep along Guide. Select the section string. the section string in this case is the generatrix, or the form that will define the shape of the part. Choose the section string. OK the dialog box to advance to the next step. Now choose the guide string. The guide string is also known as the directrix. The guide string is the path that the section will take as it sweeps. Select the guide string, select OK. We will not use any offsets because we want to make a solid component. The solid body is created, cancel the dialog box to dismiss it. Fit the view.

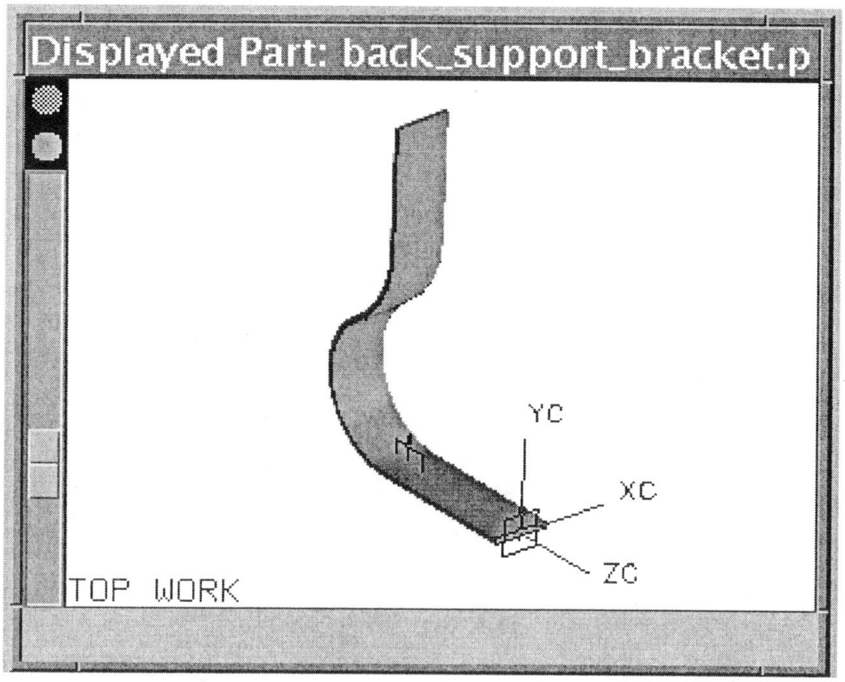

The top end of the part needs to be blended. remember that the part is 2.25" wide. Lets blend the top to be 2.25"/2. Blend all edges to be .0625.

1.6.3 Slots

Two slots need to be inserted into the top and bottom straight part to make the depth and the height of the back support adjustable. Choose Insert|Form Feature|Slot. You have a choice of 5 different slot types, we will choose rectangular, select OK. first we will choose the upper flat face of the bracket and make a slot that is .375" wide, .5"deep, (long enough to go through the entire part, and 2.5" long. The select horizontal reference dialog box appears. The horizontal reference refers to the direction that the length of the slot will go. The horizontal reference dialog will appear every time a rectangular feature is about to be created. Choose any of the edges that are parallel to the planar placement face in the long direction. Enter the parameters and choose OK.

The positioning dialog box appears. This time we will use something new. We will use line onto line. Notice that with a rectangular shape you receive a centerline from both directions. We will put this center line on the guide string to center the slot.

Choose line onto line. Again the target edge is the edge that you are measuring too. Select the sketch curve as the target line. The tool edge is the center line of the slot. We next need to position the slot up and down. What we need to do is position the arc centers of the end of the part and the arc center of the slot. Choose point onto point. The target is any arc center on the end of the support. Chose arc center. The tool edge is the arc center at the end of the slot. Choose the arc, choose arc center. The slot is positioned. Now lets create the slot for the other end of the bracket.

This slot is the same as the first one except that the length is 5". Create and position it. Center it in the placement face.

Make layer 1 the work layer and the rest of the layers invisible. Change the color to another color. Save the part as back_support_bracket.

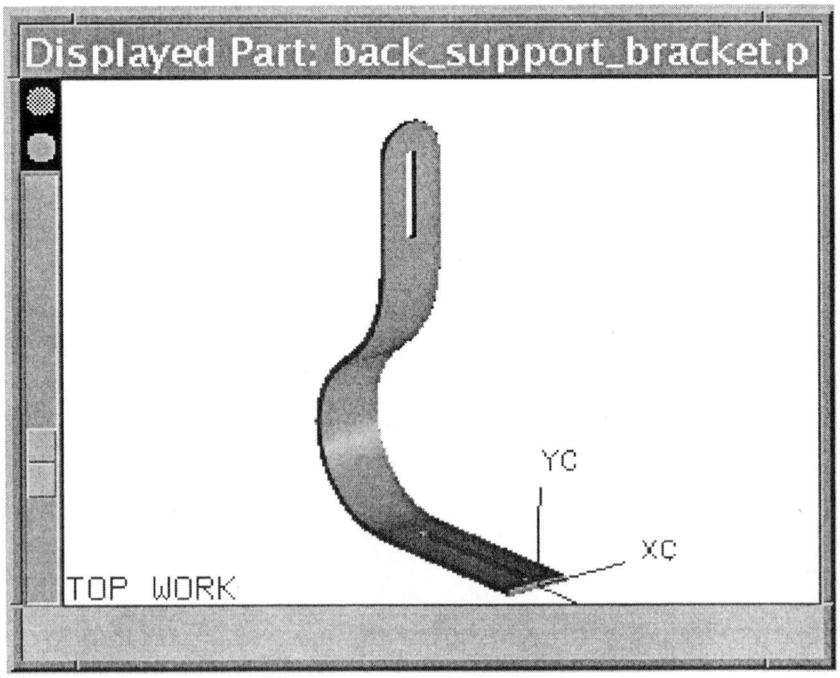

1.7 Back Plate

1.7.1 Back Plate

The back plate is a very simple part. It consists of a flat plate with a pad on it to attach to the back support and some holes in it to attach it to the back cushion.

Lets start with the plate itself. Open seed_in and enter the modeling application. Choose Insert|Form Feature|Block, choose edge lengths, corner. The dimensions are .125" in the XC direction, 10" in the YC direction, and 8" in the ZC direction. Place the block such that the WCS is in the center of the block and the back face of the block is on the YC-ZC plane. Cancel the block dialog and fit the view.

Lets blend the corners to 3". Next is the pad on the back to support the bracket. When we make a pad for the back we will want to position it in the center to help make this easier the creation of two datum planes will help. Create two relative datum planes, one from each side that cross exactly in the center of the block.

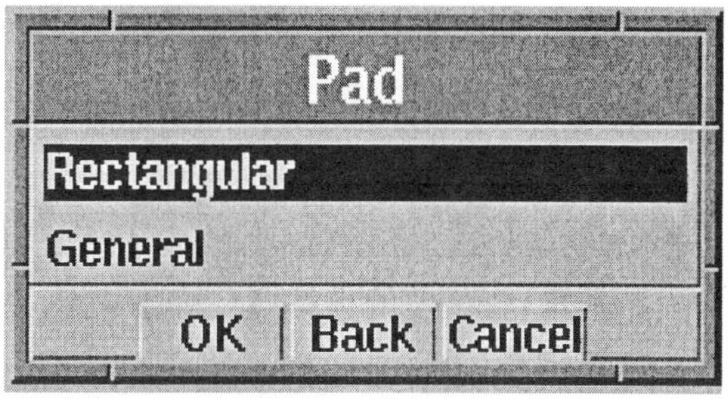

Choose Insert|Form Feature|Pad, rectangular, choose the back face for the planar placement face. choose an edge that will make the direction arrow Collinear with the YC direction. Enter the parameters. the length is 3.25", the width is 6", and the height is 1.125". As you can see you can also put in a taper angle here if you want. Lets put in 10 degrees.

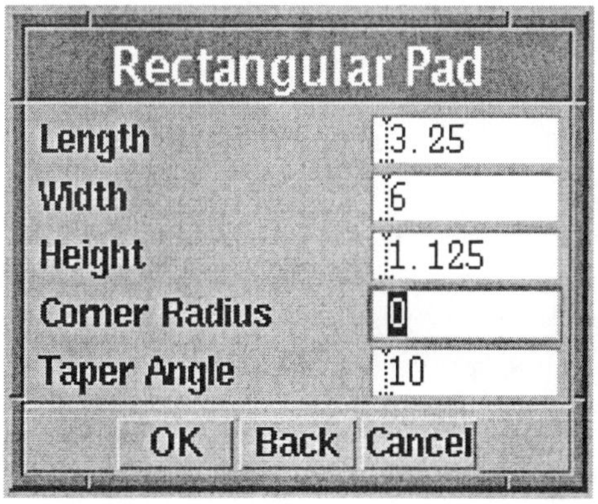

for positioning, we can use line onto line for both directions. the target edges will be the datums and the tool will be the centerlines. Choose line onto line.

We next want to put a pocket onto the pad that we just created. This pocket will act as a guide for the back support bracket. Choose Insert|Form Feature|Pocket. the Pocket dialog appears next.

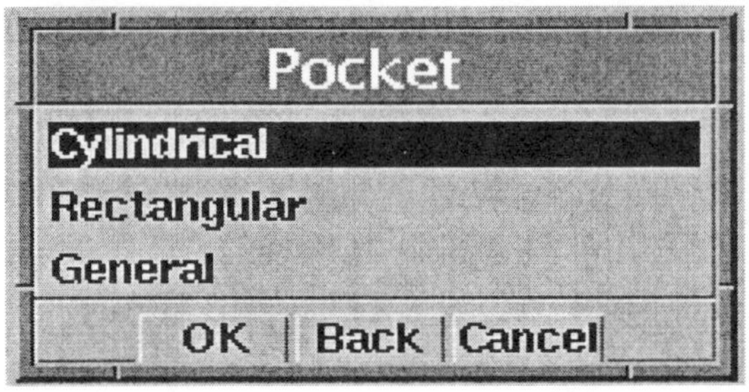

Choose rectangular, choose the pad face as the planar placement face. This time choose a long edge as the horizontal reference. The X length need to be longer than the pad so it will cut away the whole length of the pad. So lets make the length 10". The Y length needs to be the same as the width of the bracket, that was 2.25". the Z length needs to be .25, which is the thickness of the bracket.

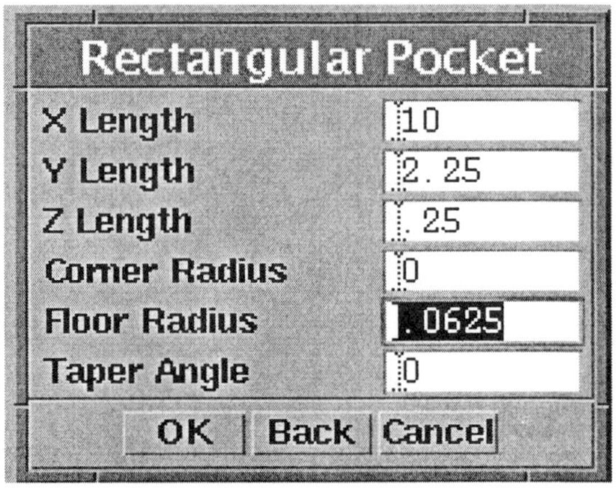

Select OK to create the pocket. For the positioning we will use the same ones that we used to position the pad, line onto line twice. Blend all edges of the plate except for the plate itself to .0625".

We now need to create a threaded hole for the adjustment knob. Choose Insert|Form Feature|Hole, simple hole, chose OK, select the face of the pocket and enter the parameters. The diameter is .375" and the depth is .75". The hole needs to be positioned in the exact center of the plate. Choose point onto line twice. Cancel the dialog and thread the hole the same way that we threaded the other holes. Now we need some thru holes in the corners to mount the plate to the seat back.

Select Insert|Form Feature|Hole. chose simple, thru, select OK. The holes ned to be positioned 1.5" from the edges. Chose point to line at a distance, or perpendicular

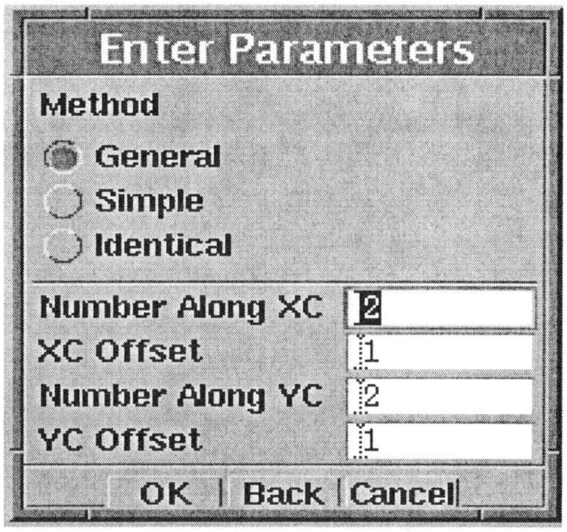

Move the WCS to the center of the hole you just made. Use dynamics to move the WCS to have the plate on the XC-YC axis. We will make a rectangular array of this hole to make exact copies in the other corners. Choose Insert|Feature Operation|Instance. The instance dialog appears. Choose rectangular. select only the last hole. Choose OK. The enter parameters dialog appears. Method can be left in the default. This is why we changed the WCS. Because we can only make an instance in the XC and YC direction. We want to end up with 4 holes total, 2 in the XC direction and 2 in the YC direction. The offset is the distance from the center of one hole to the center of the next hole. remember that the plate is 10" wide and 8" tall.Tthe offset would be 7" in the XC direction and 5" in the YC direction. select OK to preview the placement of the holes. If this is what you want select OK. Cancel the instance dialog.

Make sure that layer 1 is the work layer and all other layers are invisible. After changing the color of the part save your work as back_plate.

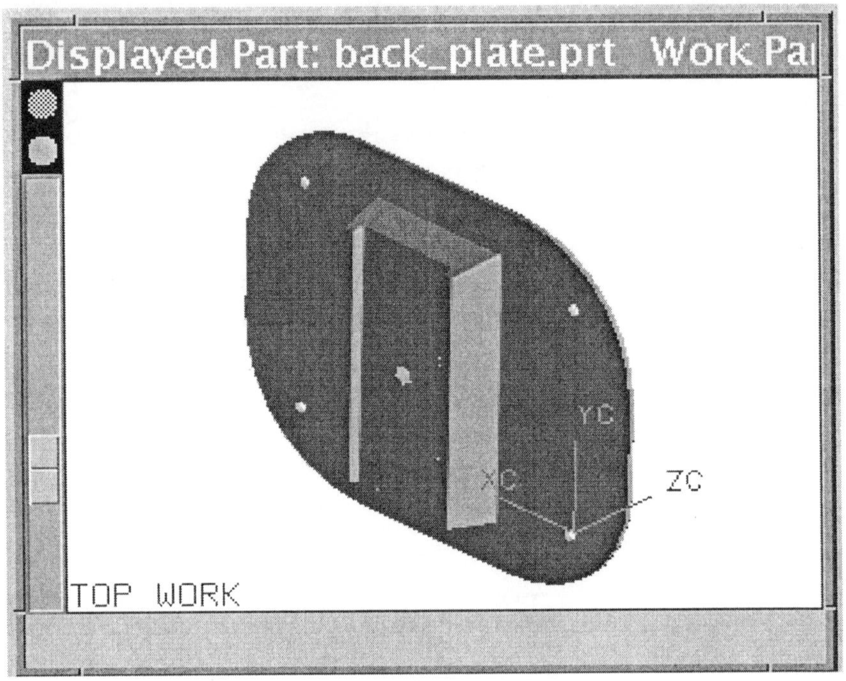

1.8 Adjustment Knobs

1.8.1 Knobs

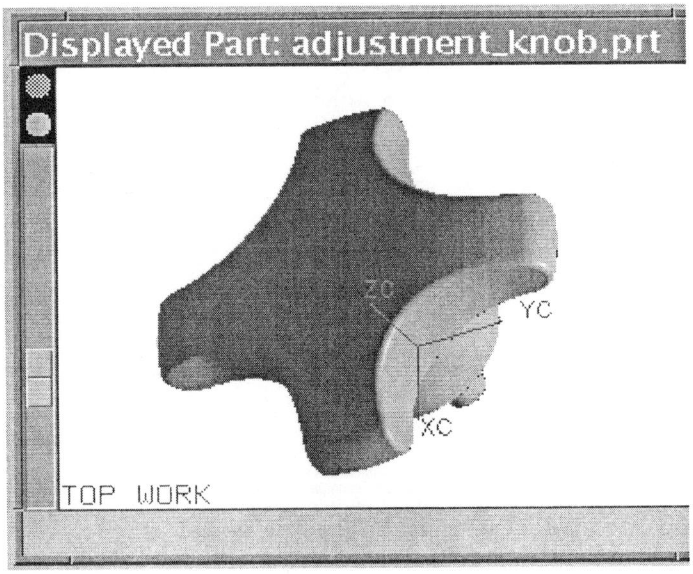

The last part of this chapter is to make the knobs that hold the back support and the seat back in position. The knobs will be the same for both spots.

Open seed_in and enter the modeling application. This part will consist of two sketches, one will be revolved and the other will be extruded. We will then apply the intersect Boolean operation to make the part. Then add a boss that we will thread, clean up the geometry with some blends.

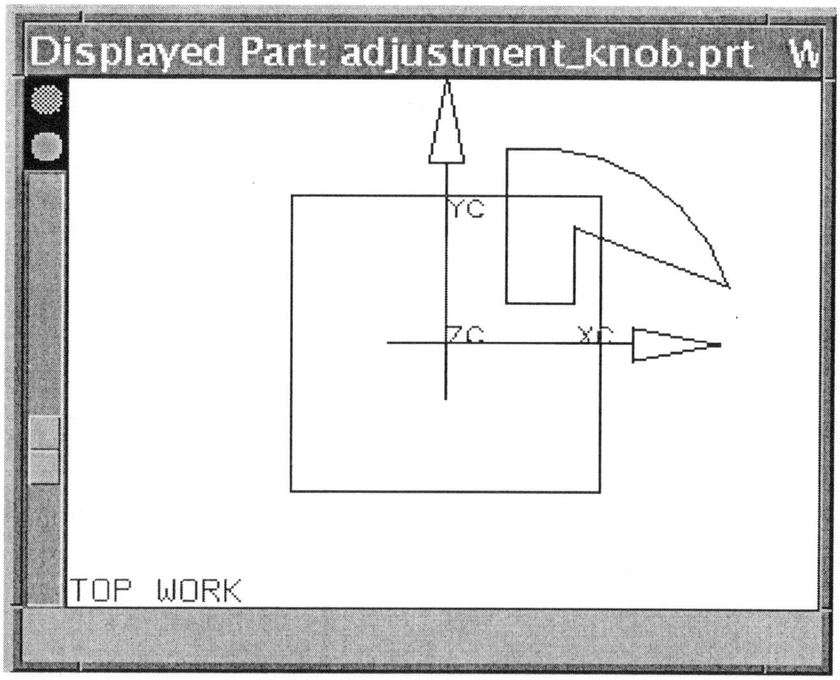

Change the work layer to 21 and then choose Insert|Sketch. Be sure to select the change view orientation radio button. Choose create sketch, change the name to s21_profile, choose the XC-ZC plane as the sketch plane, select OK. Choose basic curves, line. We will draw this line in the upper right quadrant of the sketch plane. It consists of 4 straight lines and one arc. Cancel the basic curves dialog box. We will then move the sketch into position and dimension it with constraints.

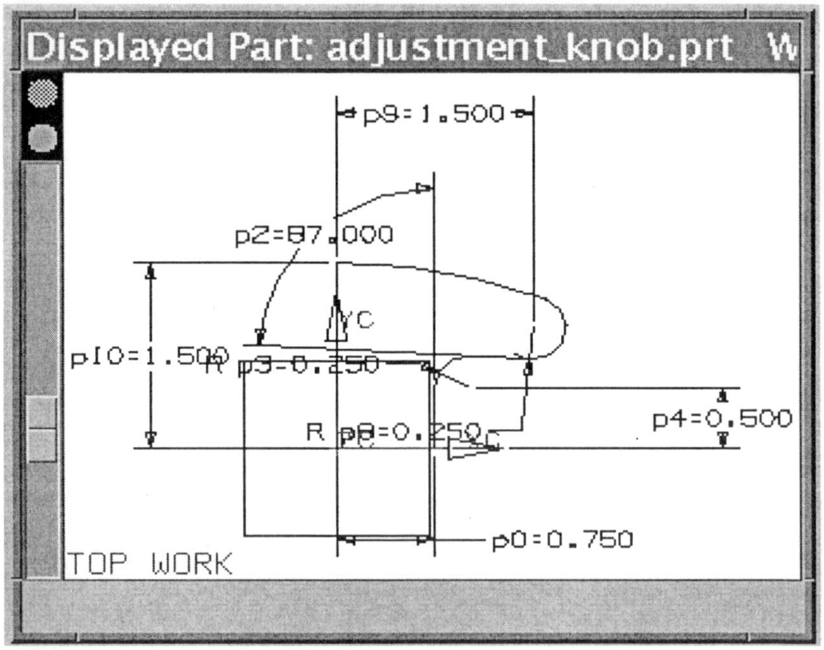

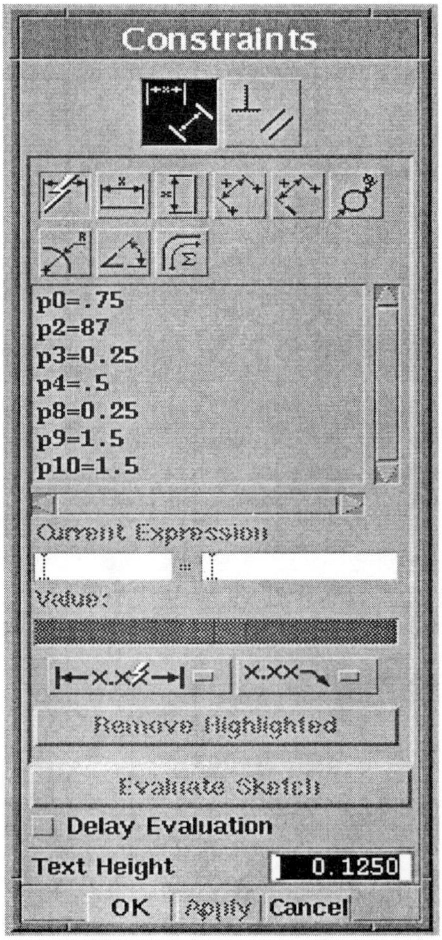

Add the following dimensional constraints. The p numbers do not have to match. That depends on the order that you choose to create the dimensions. Hint: delay evaluation may make the addition of constraints easier. Make sure that the sketch is fully constrained. You do not have to have the sketch fully constrained to model, but it makes it easier to control later. Cancel the constraints dialog and then the sketch tools dialog.

1.8.2 Revolved Body

We will now take the sketch we just made and revolve it to make a solid. Change the work layer to layer 1. Select Insert|Form Feature|Revolve. Choose the section string that you just created. The revolution method we want is axis and angle, choose OK. The revolution reference axis is the ZC datum axis. Choose the datum axis. the revolved body dialog box appears. the direction arrow that appears is for the direction of the revolution. Unigraphics uses the right hand rule for the revolution direction. If the arrow is pointing up the revolution will be counter-clockwise. The dashed arrow is the positive direction of the offset. We will not be using offset in this case so the offsets should stay at 0.

Our start angle is 0 because we want it to start at the section string. The end angle is 360 because we want it to revolve completely around forming a solid. Choose OK, a solid body is formed.

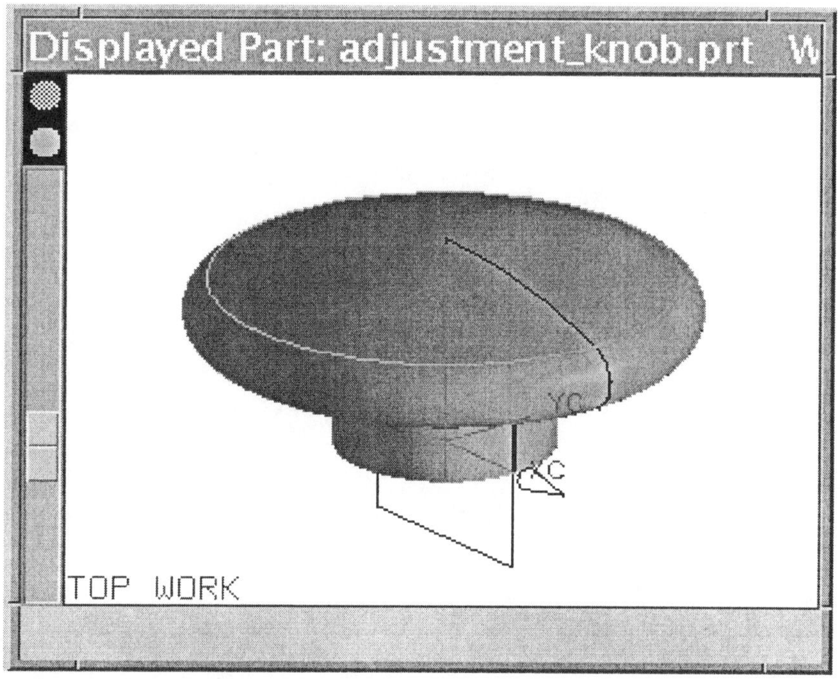

Cancel the revolved body dialog box.

We will now create a second sketch to cut away pieces of the body to make the handle of the knob.

Change the work layer to 22 for the second sketch. Choose Insert|Sketch, create. Change the name to s22_cut, and choose the XC-YC plane as the sketch plane. Choose OK, fit the view. choose basic curves. We will need 4 lines. two horizontal and two vertical.

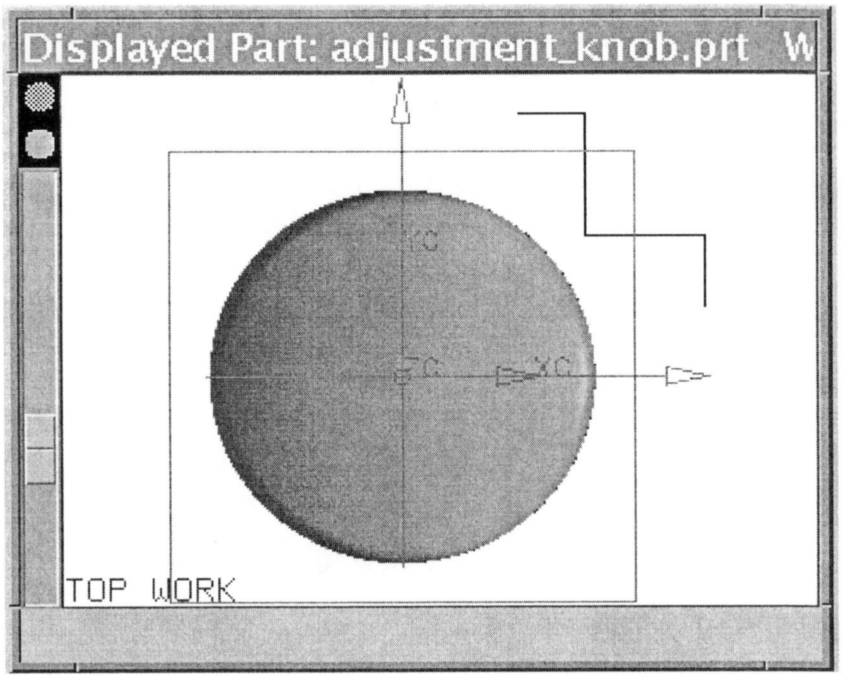

The sketch will now be constrained. The left endpoint of the upper horizontal line needs to be made a point on curve with the vertical datum axis. the bottom endpoint of the right vertical line need to be made point on curve with the horizontal datum axis.

Apply the constraints as shown. The Top horizontal line is made equal length with the bottom vertical line.

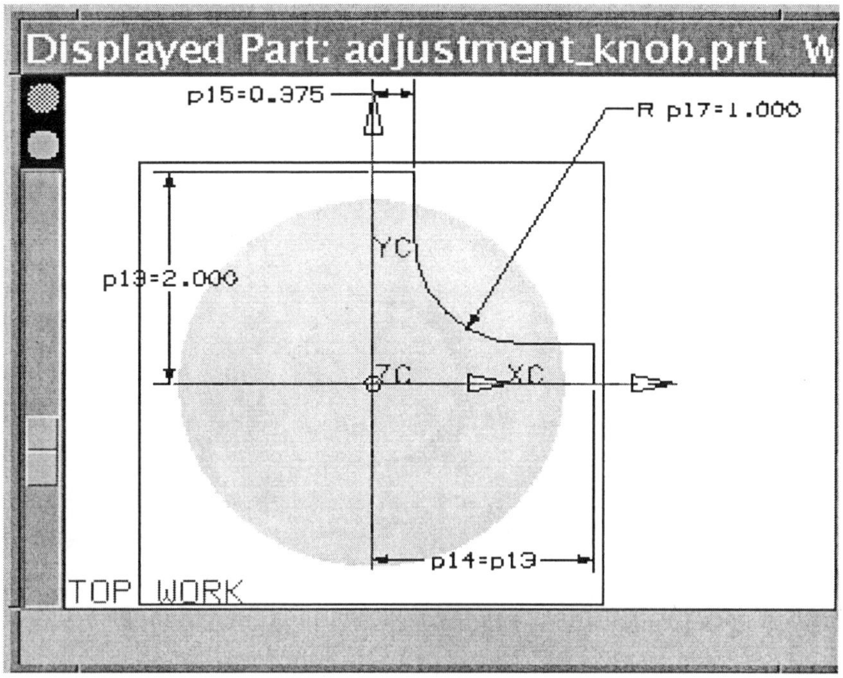

1.8.3 Mirroring a Sketch

Now we need a sketch in each quadrant the is exactly the same as the first sketch. Before we can mirror sketch geometry we need a special line to mirror about. This line is called a reference line and is easily made. Choose basic curves and create a vertical line place it collinear with the vertical datum axis, also create a horizontal line make it collinear with the horizontal datum axis. The lengths of the lines are not important. We just need a line to mirror around. It just can not be a point.

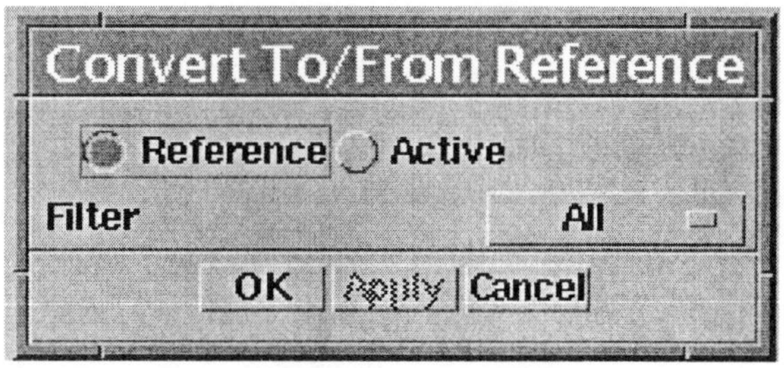

We must take these two lines that we just made and convert them to reference lines. The icon convert to/from reference icon ned to be selected. The convert to/from reference dialog appears.

In sketch tools look for the mirror icon. It is in the sketch options area.

You are asked to select the mirror centerline. choose the vertical reference line that you just created as the centerline. We will mirror one half at a time. When you choose this line, since this is the only centerline that you will use for now, select the middle mouse button to advance to the next selection step. Select the geometry that you want to mirror. It is all of the sketch that you made earlier. Choose Ok, the sketch is mirrored. Now mirror all of the sketch, (both halves of the drawing), to the bottom to complete the mirroring.

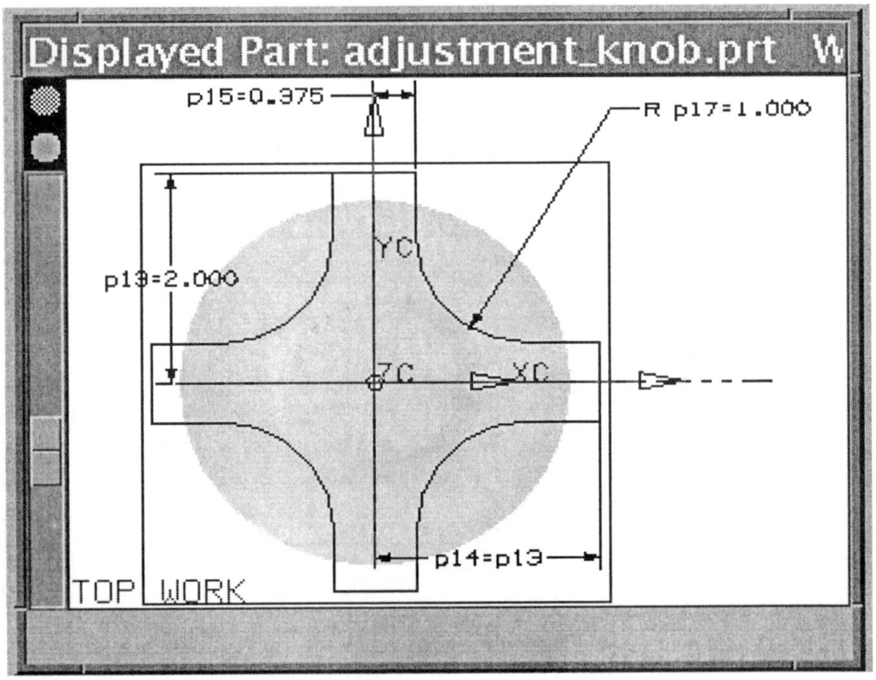

Cancel the sketch tool dialog to deactivate the sketches. The next step is to extrude this last sketch up into the part.

1.8.4 Extrude to a Face Plane

The top of the handle has a curved surface. Because of this we will have to extend the extrusion to contact all of the face. Choose Insert|Form Feature|Extrude. Select the section string, (the last sketch you did), select OK. Lets use trim to face plane. Make sure that the direction vector arrow is pointing up into the part. select OK. The trimming face dialog appears, select try both options. Select the top curved face of the revolved sketch, the one that you did first. Select OK. Accept the defaults in the extruded body dialog box. Select OK.

This time we want to use the intersect Boolean operation. We want what is left to be the common area between the two shapes. Select intersect. Cancel the extruded body dialog box.

Blend all edges to .0625". Make Layer 1 the work layer and all other layers invisible.

Now we need a boss on the bottom face of the handle to become a threaded insert. Choose Insert|Form Feature|Boss, select the bottom face as the planar placement face. The diameter will be .375", the height will be .75" with no taper angle. Select OK to form the boss, position it to the center of the handle. Cancel the boss dialog.

Thread the boss to 3/8-16 UNC. Change the color and save the part as adjustment_knob.

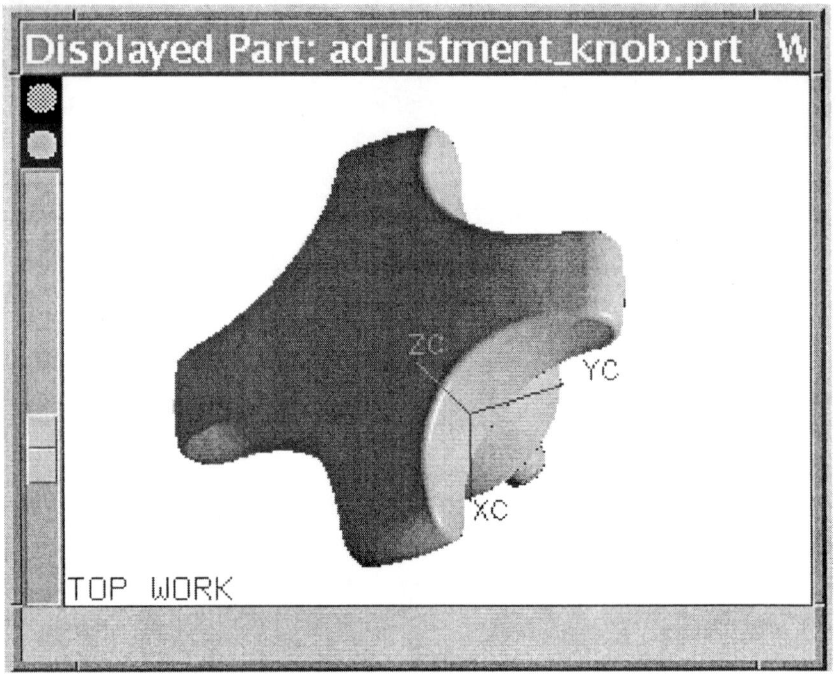
Displayed Part: adjustment_knob.prt W
ZC
YC
XC
TOP WORK

2

Unigraphics Assemblies

Overview

In this chapter we will combine all of the components that we made in chapter 1, into an assembly. Unigraphics handles assembly file in a very efficient manner. An assembly manages geometry instead of creating it. Geometry is created in the modeling application.

In this chapter you will learn how to:

- Apply master model concepts
- Create an assembly file
- Add existing components to an assembly file using the bottom-up method
- Use mating conditions
- Use the assembly navigation tool, ANT
- Create component arrays
- Create new components
- Add new component to an assembly file using the top-down method
- Create an assembly drawing with an exploded view
- Use the wave geometry linker to establish geometric relationships between assembly components
- Create exploded assembly views

2.1.1 Master Model Concepts

Unigraphics uses the assembly modeler approach to construct assembly files. This means that all components are actually links to the component parts. This is a very important concept because your memory requirements are reduced, your assemblies are automatically updated if a master part is changed, and most importantly, there is only one copy of your geometry, so edits and changes that are made to the master part are changed in all assemblies that use that part.

An assembly is a file that contains objects, or components. A sub-assembly is an assembly file that contains component part of its own.

A component object is a link that references back to the master component. A component object contains information like; name, information about how the part is displayed, what layer its on, what mating conditions are assigned to it, etc.

It is important to remember that a link is just that, a link to the original part. When an assembly file is saved you save the link not the part. Also if you had a simple assembly with 4 components you would have 5 part files. One for the assembly and four for the components.

The concept of assembly modeling leads to the concept of a master model. With a master model diverse yet dependent design processes can take place during product development. Each application uses a separate assembly file. You could have an assembly file for the manufacturing application, one for finite element analysis, FEA, one for industrial design, etc. When the master part is modified the other applications will automatically modify with little or no loss of associativity. Also, while using the master model technique, the design intent of the designer can be maintained through model protection of the master model.

This master model example eliminates the duplication of model geometry and more importantly it enables concurrent engineering. Meaning that different users can work with separate files simultaneously while referencing the same data.

2.1.2 Assembly Concepts

There are two basic ways to create an assembly. Top-Down and Bottom-Up assembly modeling.

Top-Down assembly modeling is used when you have a concept in mind but no, (or not all of the), components.

Bottom-Up assembly modeling is used when you have the components and are creating the assembly. Components are added as you move up the assembly level hierarchy.

Most of time you will use a combination of these two methods. This happens when you have some of the components that you can use from an existing product but also need some new parts for a new product.

2.1.3 Load Options

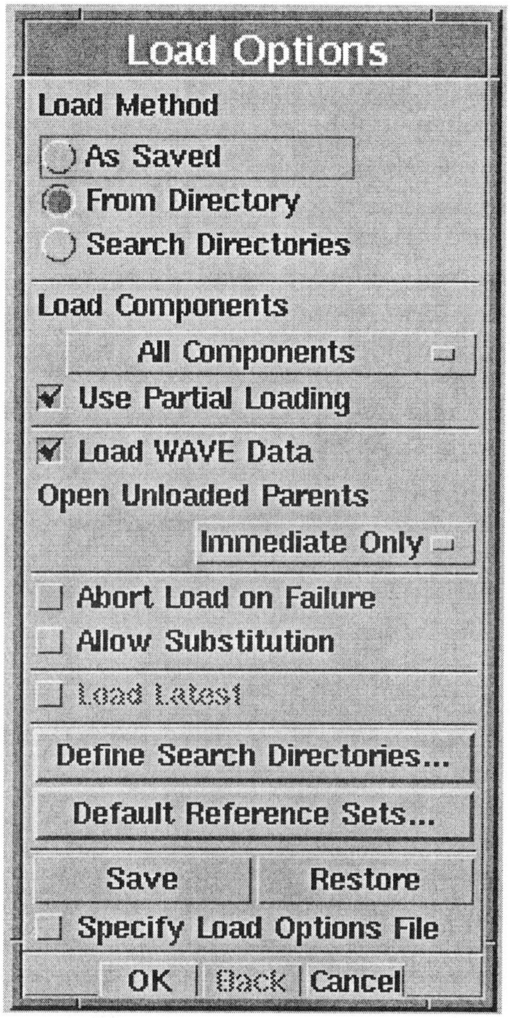

With our ability to use geometry from other places, it becomes important to understand that files may in reside anywhere that you have access to. If you are using components from different designers that are part of your assembly, Unigraphics has to be able to find them. The default is to look in your directory. We have the ability to tell Unigraphics to look everywhere or at specific locations or directories.

The load options dialog is accessed by selecting File|Options|Load Options. Later when we start loading assembly files and you run into problems with loading parts that the system can't find, this is almost certainly the problem. The load method option will need to be set to as saved.

2.1.4 Assembly Navigation Tool, ANT

Just as with modeling, we also have a tool at our disposal to help us easily manage the many aspects of assembly modeling. The ANT provides a graphical display of the assembly structure, or the assembly tree. The ANT provides another way to work with components and perform assembly operations.

The ANT is accessed by selecting View|Assembly Navigator. The default configuration is very adequate for the needs of this course.

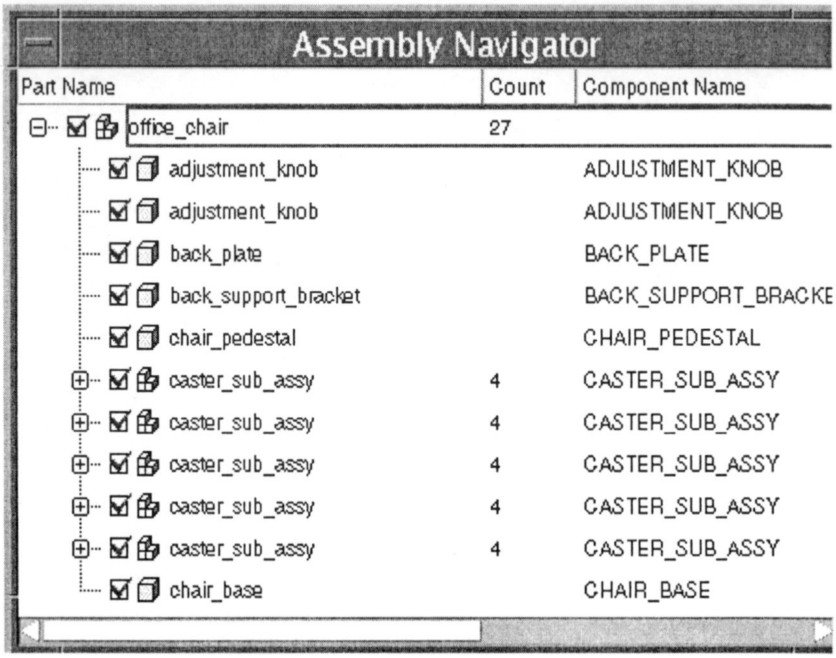

This is what the assembly navigator will look like for our chair project. As you can see, a + sign means that there are components in a sub-assembly. A red check means that the part is active and a box with no check mark means that the component is suppressed. In the example of the caster_sub_assy sub-assembly there are 3 separate component parts but the count is 4. This is because the other file is the assembly file containing the links only.

2.2 Assembling the Chair

2.2.1 Application Assemblies

Starting with V 17, the assemblies function is an application. Select Application | Assemblies. You can create assemblies in the gateway or the modeling application.

We will start by using the bottom up assembly method. We will be using the components that we made in chapter 1.

2.2.2 The Caster

Lets start with the caster. It will be a simple assembly that will become a sub-assembly. Open seed_in, enter the assemblies application. Choose File | Save As, and save the file as caster_sub_assy. For now this file is the same as all other part files. Unigraphics makes no distinction at this point. Choose Assemblies|Components|Add Existing. As soon as you add a component to the file it becomes an assembly file. It still has the .prt file extension. The select part dialog box appears.

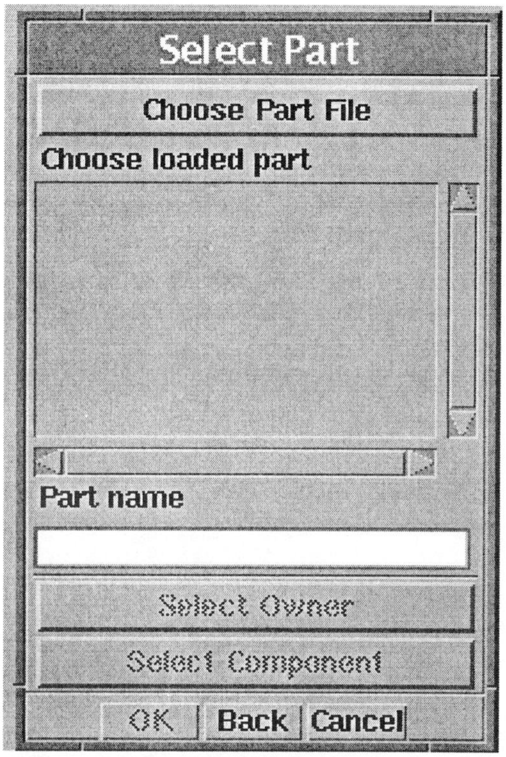

If you have been working on some components a list of those components will be seen in the choose loaded part list area. If not, select choose part file. The part name dialog appears to select the component that you wish to add to the assembly file.

The first part chosen makes a difference. You should think about a part called the base part. This base part is the part that all other parts will be added to. Lets choose the caster_center_stem component. The add existing part dialog appears. From here we can see the component name that we are about to bring into the assembly. We can change how many we add at one time. We can change the positioning method, and how the layers are handled when the part is brought in.

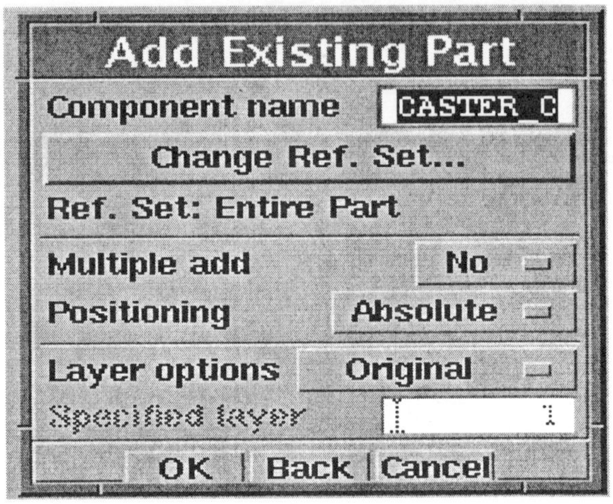

We will use the settings in this box as shown. Since this is the first part, we will place it at the WCS. Select OK. The point constructor dialog box appears lets place it at o,o,o. The part is brought into the assembly file. This is when the .prt file changes from a normal part file to an assembly file.

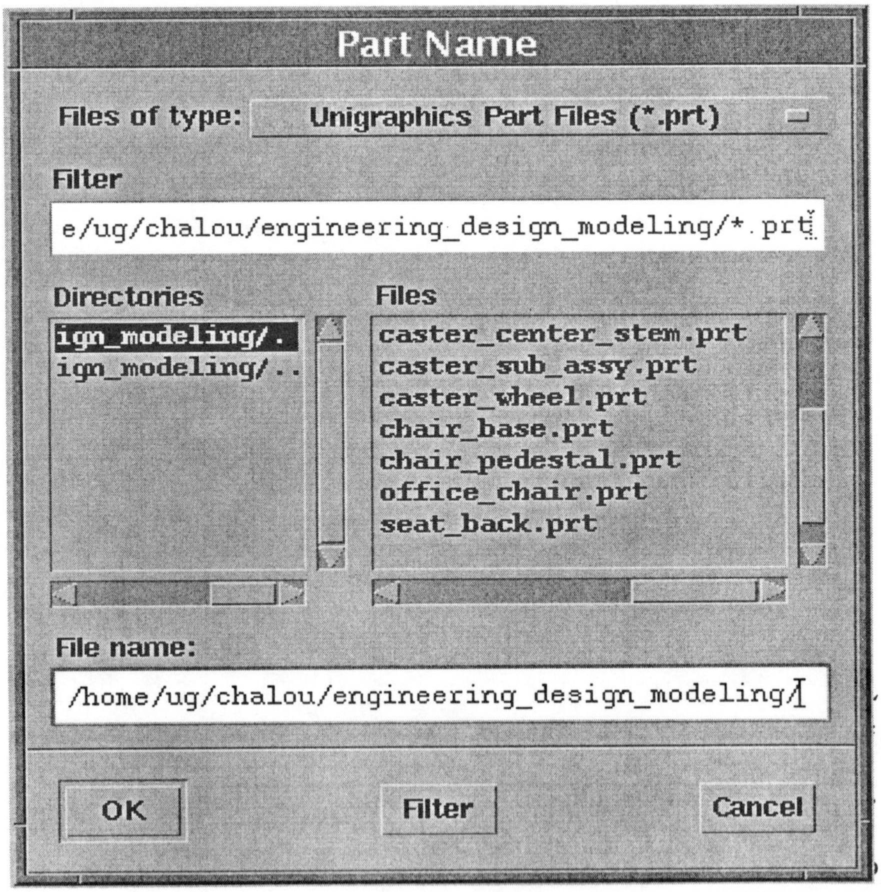

ing in the next part. Lets choose the caster_wheel and bring it in the same way, but this time change the positioning method to mate. When the point constructor dialog appears, move your cursor into the graphics area and select a spot for the WCS of the wheel to be placed.

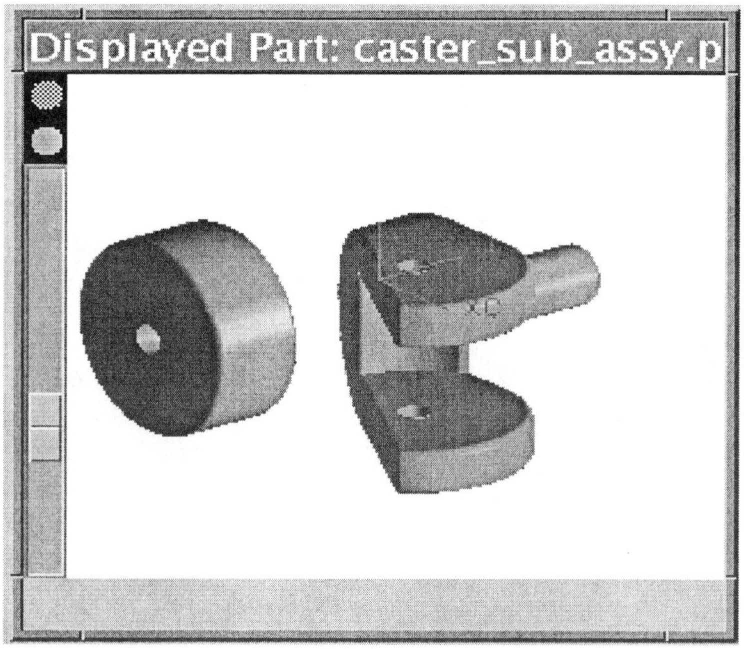

2.2.3 Mating Conditions

Mating conditions are ways to have the assembly behave as you would expect it to behave. When mating conditions are applied to assembly component parts that have relationships to one another, these parts will always have these relationships maintained. If a shaft needs to be always centered in a hole, we can assign a center mating condition to the shaft and hole and they will remain that way until this mating condition is removed.

The mating condition dialog box appears because you have selected mate as the positioning method when the new part was brought into the assembly. The cue line is asking you to select the from component. The from component is the component that you will be moving to it's new position. We will be moving the wheel to the caster shaft. Next we have to decide what type of mating condition to apply first.

As you can see, the wheel is 90 degrees to the caster center stem and needs to be rotated. In this case it really doesn't matter which mating condition is applied first because both sides of the wheel are the same. Lets find the icon for center.

When you choose center the center objects selection area highlights. If you look at the pull down menu you can see that there are 4 different ways to center objects. 1 to 1 is when you have one cylindrical object centered on one cylindrical object. 1 to 2 is when you have one cylindrical object centered between two flat faces. 2 to 1 is when you have two faces centered in one cylindrical shape. 2 to 2 is when you have 2 faces centered on, or in, two faces.

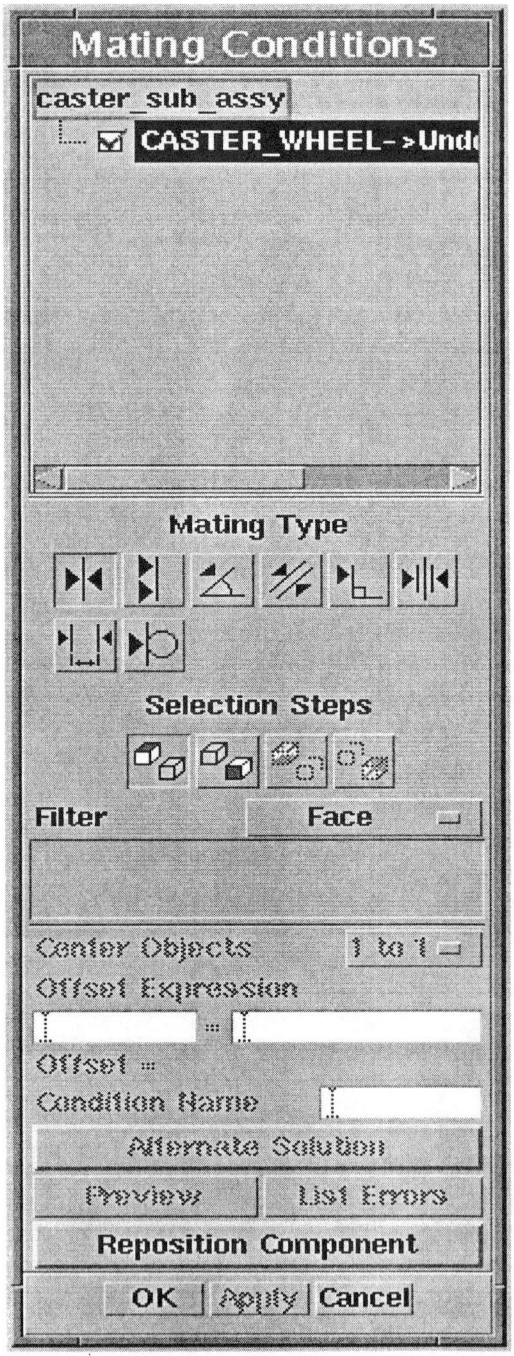

For this first step we will choose 1 to 1. Choose the inside face of the center hole of the wheel. We will be moving the wheel to the caster center stem. We want to center the hole with the hole in the caster stem. Choose one of the holes as the to component. Direction vector arrows appear indicating what degree of freedom will remain when this condition is applied. Choose apply. The wheel centers on the hole.

The freedom arrows indicate that the wheel can still rotate and move vertically. Choose reposition component and move the component around. You can see that you can move it up and down and rotate it but you can not move it out of the center line of the caster stem component hole.

Leave the wheel above the caster stem, it will make it easier to see the next faces. We now want to make sure that the wheel can't move in the vertical direction, or along the centerline axis of the hole. What we want is for the wheel to be centered in the slot in the center of the caster stem.

Choose center and in the center object area, choose 2 to 2. Pay close attention to the cue line. Select the first from component. It will be the top face of the wheel. Choose the corresponding face on the inside of the caster stem as the first to component face. The second from face is the opposite side face of the wheel and the second to face is the other inside face of the slot.

Notice that the only freedom of movement that remains is for the wheel to be able to rotate. Choose apply. Choose OK to dismiss the mating condition dialog box.

The next part to add is the axle for the caster. Choose Assemblies|Components|Add Existing. Choose part file, caster_axle, OK. We will also want to mate this part to the caster_stem.

Whenever possible always move the components to the base part, and keep the base part the base part whenever possible. Position the axle the same way you positioned the wheel. Center the axle into the hole of the caster stem. Choose apply. Reposition the axle out of the part to make selection easier.

Choose center, 2 to 2, and center the axle in the part. Choose apply, choose OK to cancel the dialog box. This is our first assembly.

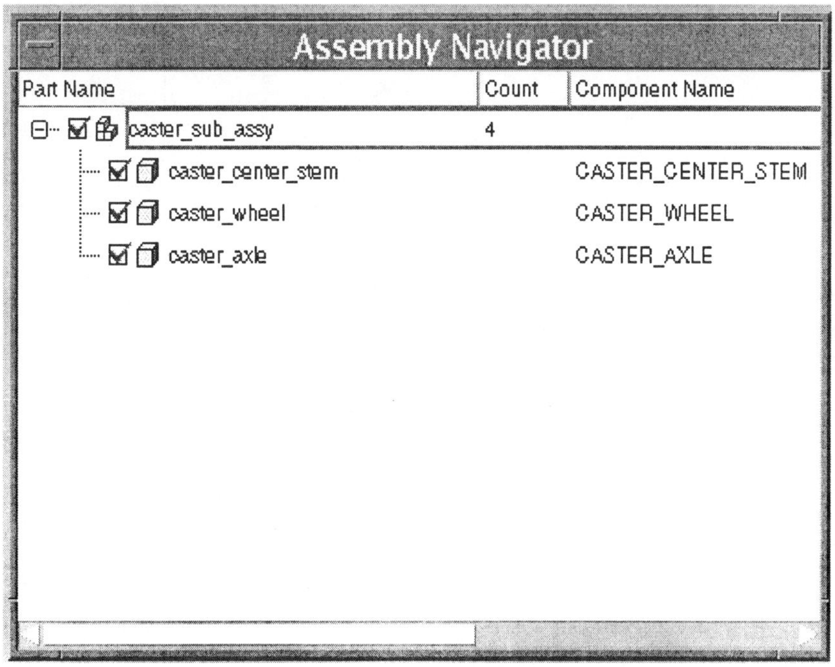

This is what the assembly navigator should look like at this point. Remember that there are four components associated with the part file but only three component parts.

What is really in this file? Remember that they are only links to the components that are really someplace else. Be sure to save this part as caster_sub_assy.

2.2.4 The Seat Base

Open seed_in, we will now create the main assembly file. Save this file as office_chair. Choose Assemblies | Components | Add Existing. The part that makes the most sense to use as the base part is the base. Lets add the chair_base to 0,0,0 and use absolute as the positioning method. After every addition you must cancel the add existing part dialog and reopen it to add the next part.

2.2.5 The Center Pedestal

Now we can add the center pedestal and assign some mating conditions to the parts. Choose Assemblies | Components | Add Existing. Choose part file, chair_pedestal, make sure that you change the positioning method to mate. Select OK. Indicate the position in the graphics area. This time lets start with a different mating condition. Where the pedestal fits into the base, two surfaces touch and when the chair is rotated these two surfaces are in contact. What we want is for these two surfaces to always be in contact. Find the icon for mate in the mating type area.

Remember that we are moving the pedestal to the base so the pedestal is the from component and the base is the to component. For the from component lets choose the face at the bottom of the boss, that is the larger face on the center stem. The to component will be the top face of the base. Choose apply. What happened is that the two faces are now in alignment, if you rotate the parts around you can see that the two faces are now on the same plane, and the volume of the part is going in the opposite direction. On the other hand align is just the opposite. with align the faces are on the same plane and the volume of the parts are on the same side. The freedom arrows indicate that the part can still rotate and move in the XC and YC direction but not in the ZC direction.

We will now center the shaft into the hole of the base. Choose center, 1 to 1. Choose the small cylinder of the pedestal as the from component and the hole in the base a s the to component. Choose apply. The only freedom arrows left indicate that only rotation is allowed.

This is what we should now have and the ANT should look like this.

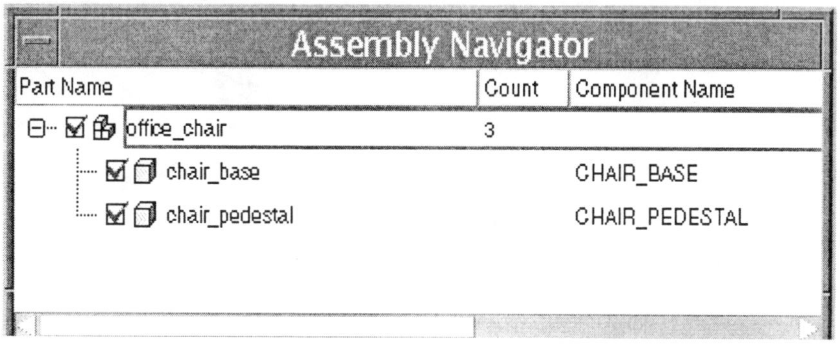

2.2.6 The Back support

The next part to add is the back support. Choose Assemblies | Components | Add Existing. Choose back_support_bracket. Use mate as the positioning method and select OK. Position the bracket in the graphics area. The first thing to do is to put the surfaces that will touch on the same plane. Choose mate, start with the top of the long arm, (the arm that attaches to the chair pedestal), of the bracket as the from component. Choose apply.

The next operation is to line up the holes that secure the bracket. The bracket has to be able to slide back and forth for adjustment. so the holes have to line up with the slot. Lets use center, 2 to 1. The first from component will be one face of the slot, the to component will be one of the holes in the pedestal, and the second from will be the opposite slot face. Choose apply. Choose reposition and position the bracket in an approximate position where you can see both holes. You can see that you need one more mating condition applied. If you could make a side of one of the parts parallel with a side from the other part that would stop that rotation. Choose parallel, choose a from edge and a to edge then choose apply. You can change the filter to help make choosing an edge easier. Choose OK to cancel the mating conditions dialog box.

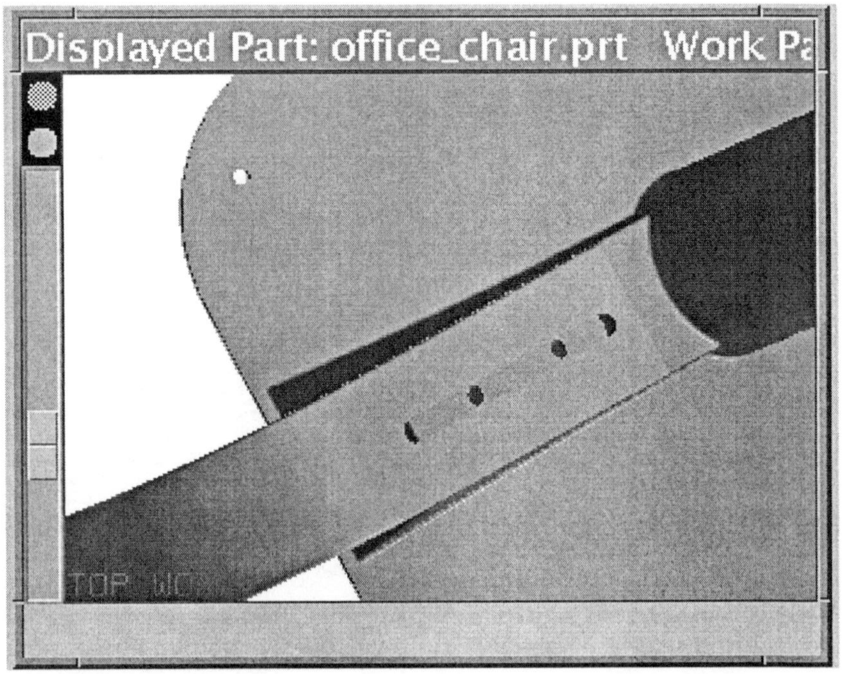

This is what we have so far and the ANT will look like this.

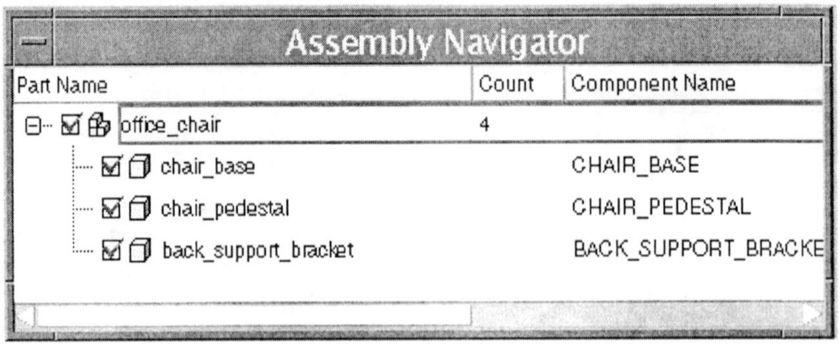

2.2.7 The Back Plate

By now it is really starting to look like an office chair. We will now add the back plate to the bracket that we just added. Choose Assemblies|Components|Add Existing. Choose back_plate. Use mate as the positioning method and select OK. Choose a spot in the graphics area to place the back plate and select OK. The mating constraints dialog box appears.

The slot in the back pad of the back plate needs to be lined up with the back bracket. Lets do that first. Choose center and select 2 to 2, since we have two faces that have to line up with 2 faces. Select the inside side face of the back plate pad as the first from face and the outside face of the bracket as the first to face, then select the opposite faces for the second to and from faces. Select apply. Now mate the back face of the back plate, (the one with the hole), to the front of the bracket. Reposition the back plate to the proper spot. This can be tricky because the bracket can also still move. Cancel the add existing part dialog box.

And the ANT look like this.

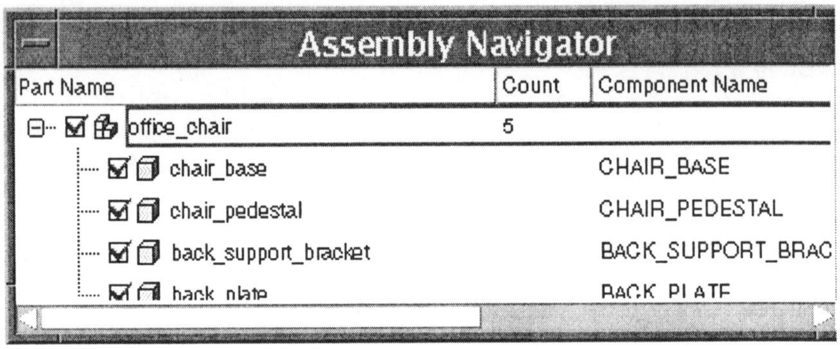

2.2.8 The Adjustment Knobs

We now have to have a way to attach and secure the back plate to the bracket and the bracket to the seat. Choose Assemblies|Components|Add Existing. Choose adjustment_knob. We have the ability to change the multiple add box from no to array or chain. In this case because the knobs are not in a particular pattern will leave this at no. Choose mate to get the knob in the proper orientation. Lets put this knob on the back first. Choose the inside face of the knob and the bracket face near the top, select apply. Now center the threaded bolt portion to the threaded hole. Select OK to get back to the add existing part dialog. Since we are adding the same part select OK. Put another knob on the bottom of the seat. Cancel the add existing part dialog.

This is what the assembly looks like.

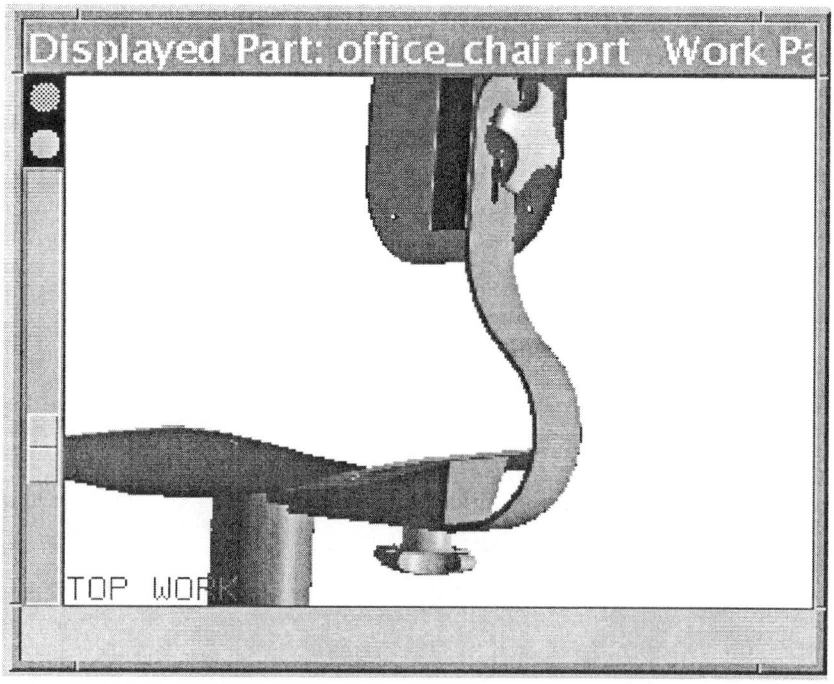

And this is what the ANT look like. Notice that there is a 7 count but only 6 components.

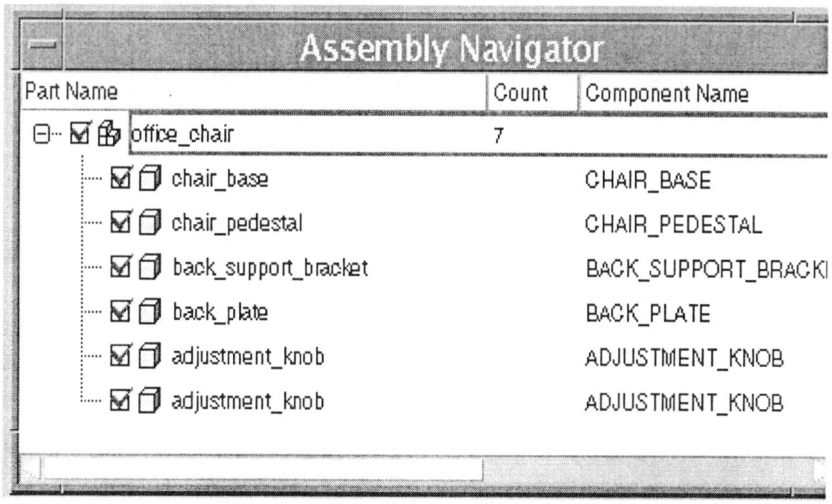

2.2.9 The Casters

This is where we put in the casters. The casters are in a pattern so we can use a different approach to position them. There are 5 casters needed for this assembly. Choose Assemblies | Components | Add Existing. Remember that the casters are in fact an assembly themselves. When an assembly is added to another assembly they become a sub-assembly. Choose caster_sub_assy. You can leave multiple add at no.

Mate the caster to one of the holes in the chair base. cancel the add existing part dialog. Choose Assemblies|Components|Create Array.

Choose caster_sub_assy. Choose circular from the create component array dialog, choose OK. We can use any circular face from the central pedestal as the center of our array. Choose a face, the total number and angle box becomes active. We need 5 total instances and the angle can be inserted as 360/5 for equal spacing.

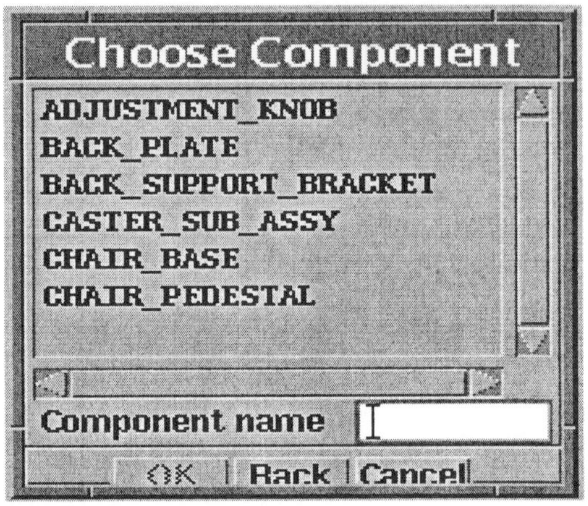

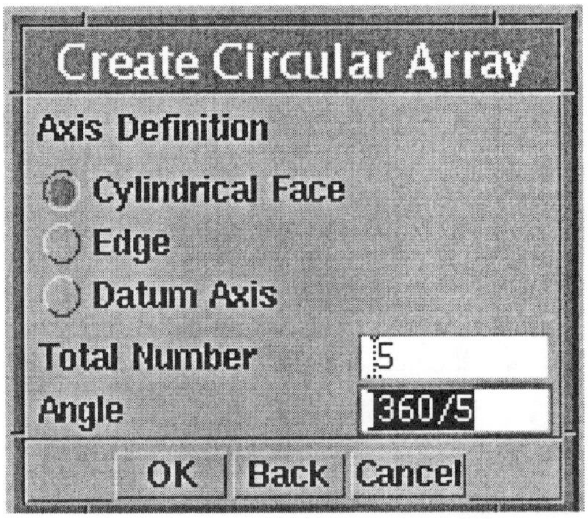

Choose OK and the casters are added to the file and to the ANT

And this is what the ANT looks like so far.

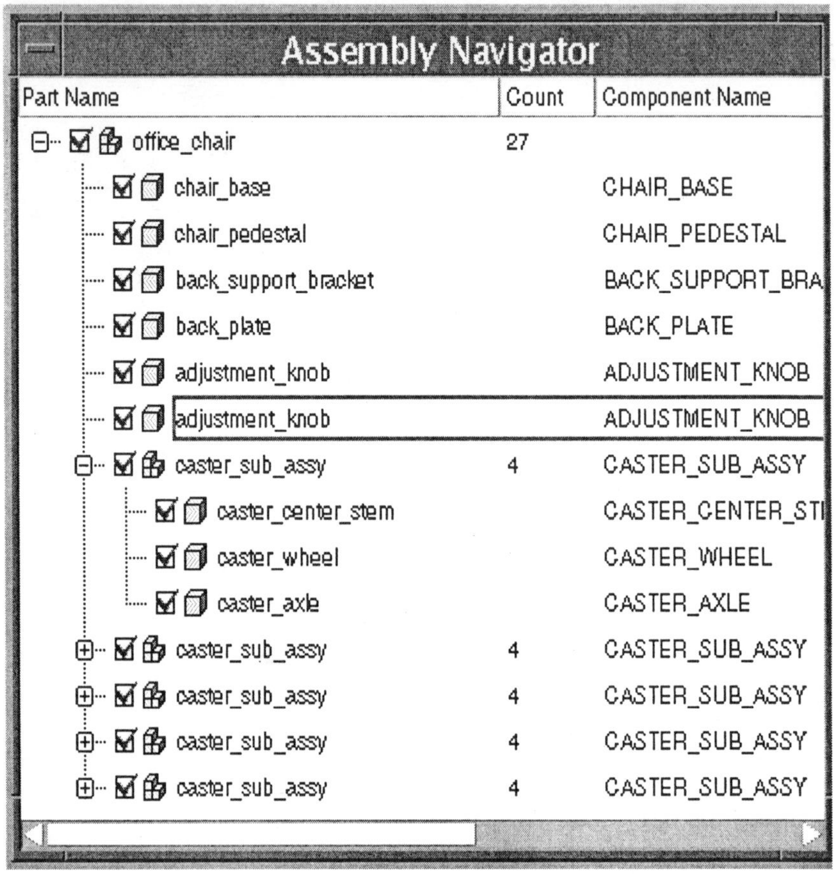

2.3 Top-Down and Bottom-Up Modeling

2.3.1 Top-Down Modeling

Up to now we have been using existing parts for our chair. This is called bottom-up assembly modeling. What do we do if we don't have a part designed yet for our assembly? We know that we will design a part at some point but we don't have it yet. We need to have a place to put the geometry once it is produced. To do the we must create a place holder for our geometry. Choose Assemblies|Components|Create New. The class selection dialog appears. It allows us to select existing geometry to change, or edit, to make something a little bit different, or we can create something entirely new. In this case we can just OK the dialog box, because we are creating something that doesn't exist yet.

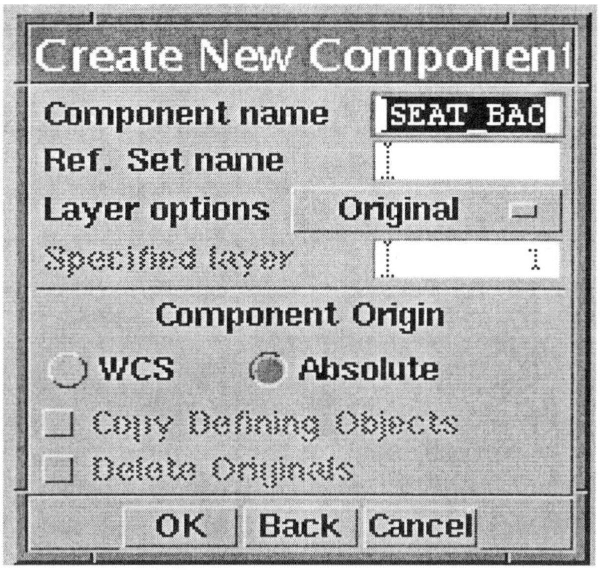

The Part name dialog appears this is where you assign a name to your new part. We will be creating place holders for the parts that we will be making in the freeform chapter. We will be making the seat cushions. For this first file, lets call it seat_back_cushion. select OK and the create new component dialog box appears.

We can accept the defaults. select OK, and notice where the seat_back_cushion link went in the ANT. It went into the higher level assembly because that is the work part. If you wanted the part to be added to a different place, like into a sub-assembly, that would have to be the work part. This can all be done through the ANT. Create another component and name it seat_bottom_cushion. Cancel the class selection dialog.

This is what the ANT looks like now.

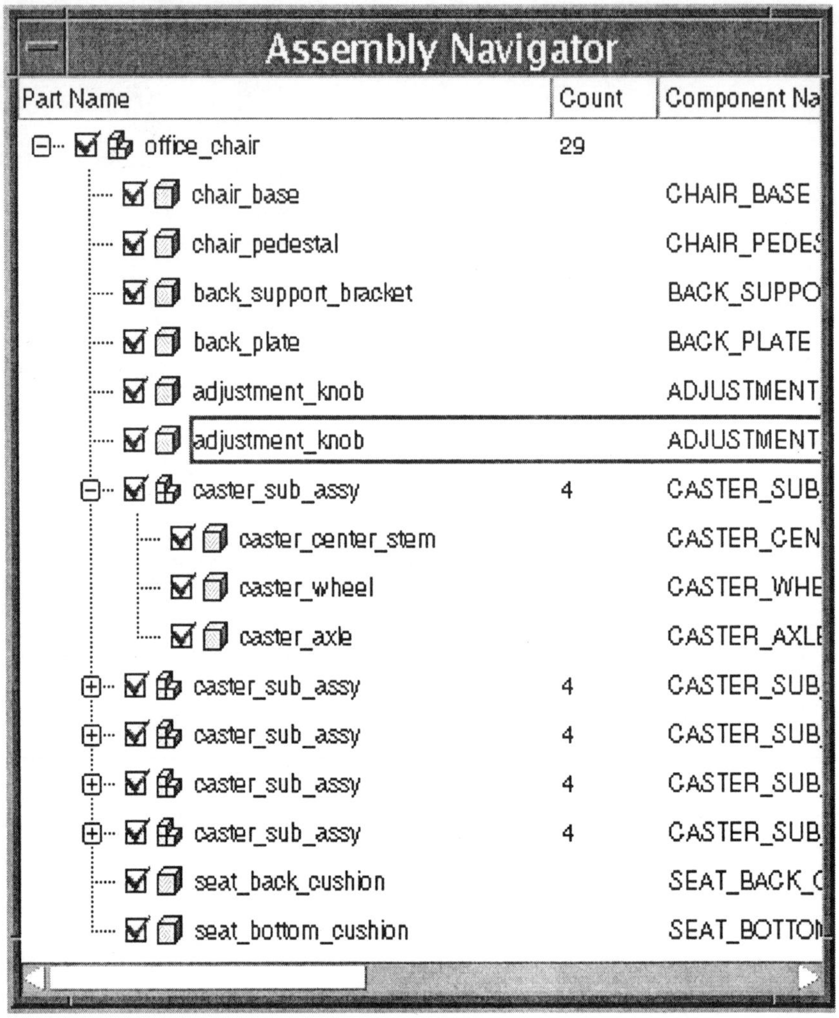

You now have some part files that have no geometry associated with them yet, but they are part of the ANT and part of the assembly. You can open and work on these parts just like normal.

2.4 Wave Geometry Linker

2.4.1 Wave Geometry Linker

With Wave Geometry Linker we can associate geometry from one part to another part, that is the work part, in an assembly. You can choose different types of geometry to link,

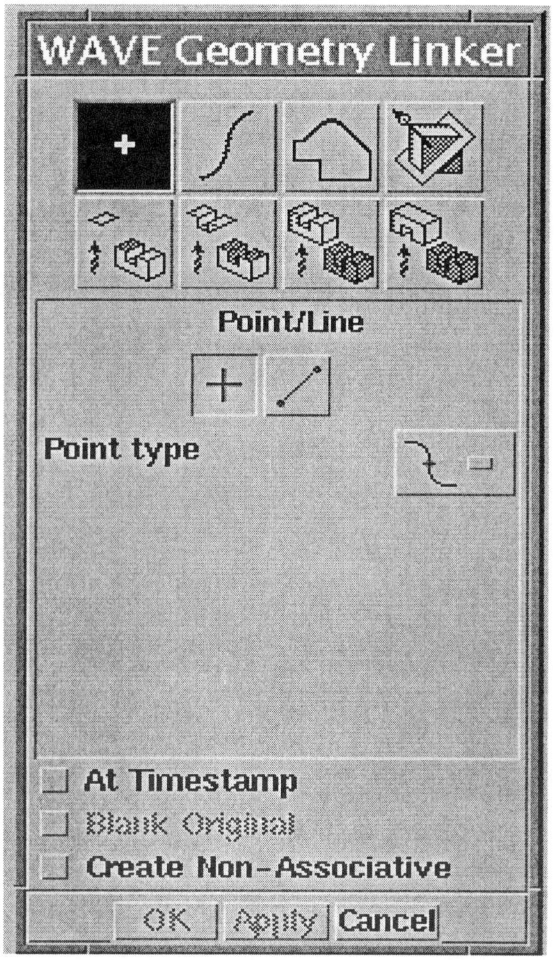

With the geometry linker the geometry that you link is associative to the geometry that you are linking it to. If the original geometry changes then the associated geometry changes. These components are driven by the original part. Lets open office_chair and enter the modeling application. Make sure that application assemblies is selected, if it is not already. We are getting ready to model the seat cushions. Since the cushions have to fit on the back plate and the seat bottom plate, we will use this geometry as a starting point.

Using the ANT, make seat_back_cushion the work part. notice how everything in the graphics area grays out. Remember that there is nothing in the seat_back_cushion file yet.

Wave geometry linker is accessed by choosing assemblies|wave geometry linker. You can see from the dialog box that you can choose many different types of geometry. Points, curves, sketches, datum planes, faces and bodies. The at timestamp button lets you specify where the feature is placed in the ANT feature list. With this button off, any feature added later to the original geometry will be associated in the new part. When turned on, any new feature will not affect the linked geometry.

When you use the geometry linker you create a copy of the geometry that you are linking. When you do this it places a copy of the geometry at exactly the same place. This can sometimes become confusing when trying to select geometry, you don't really know what you are choosing. sometimes blank original will help in the selection process.

2.4.2 The Back Cushion

We will start using the chair back as a guide for the seat back cushion that we will model in the freeform chapter. For now we want to have the shape of the seat back plate control the shape of the seat cushion. Choose assemblies|wave geometry linker. We want to select the front face and all the edge faces as our controlling geometry. Choose the region button. The cue line asks for the seed face for the region. The seed faces are all the faces to be included. Lets choose the plate front face, and all the plate edge faces. Select OK. The cue line then asks for a boundry face the boundry face is the back face of the plate.

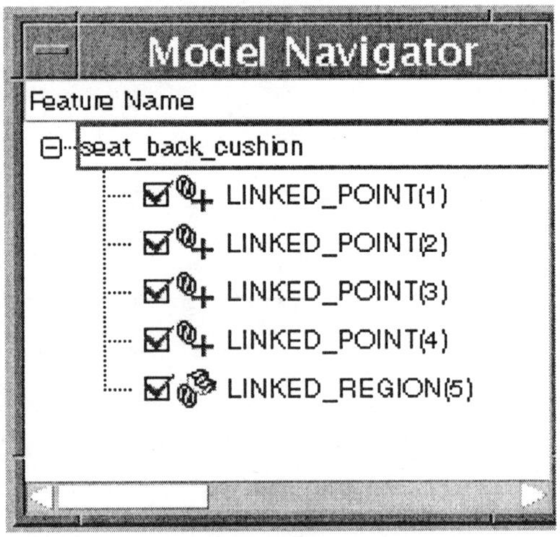

We will also want the position of the mounting holes to be associative to the seat cushion. Change to point and select all four arcs to fine the centers of the holes. Be sure in this case to select blank original. Select OK. If necessary select cancel to dismiss the dialog box.

The MNT will look like this for the work part. The original geometry is now blanked in the work part view. Using the ANT make the seat_back_cushion the displayed part. The displayed part and the work part are now only what is seen in the graphics area. The geometry that is now visible is what will be used to define the shape of the seat cushion. All it is right now is a sheet. A sheet has no volume. We now have to turn it into a solid.

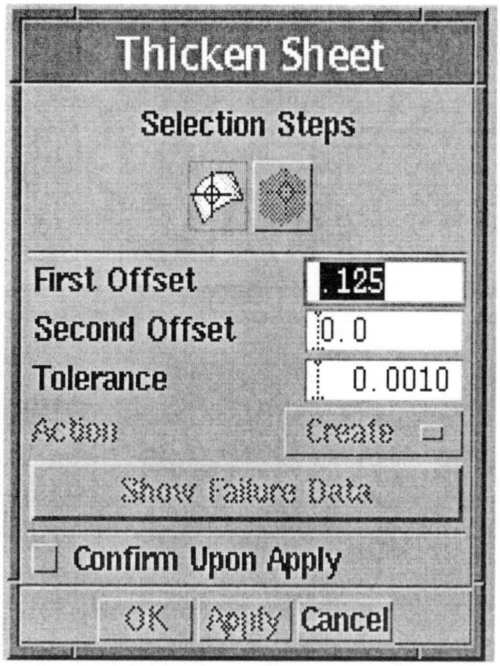

Choose Insert|Form Feature|Thicken Sheet. Choose the shape and thicken it to .125”. Select the sheet and make sure that the direction arrow is pointing out. Select OK.

If you shade the view you will see that the original geometry is visible along with the solid body. Lets blank the original geometry. Select Edit|Blank|Blank, choose the sheet body, select OK. Shade the view and you will see the part. The hole centers are still in red, we will use them later to make the threaded inserts to attach the seat cushion to the plate.

We now need to offset the side faces to extend them out a distance. Choose Insert|Feature Operation|Offset Face. We will offset the face 2” away from the center. Select OK.

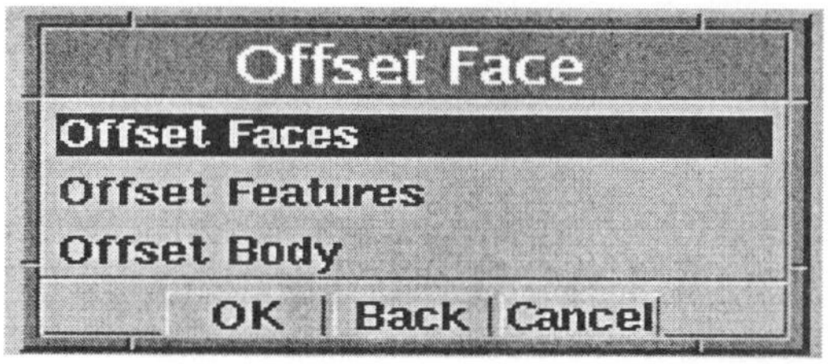

The offset faces dialog box appears. Choose OK. Choose all of the outside faces.

Select OK. The solid body is extended to the proper dimensions.

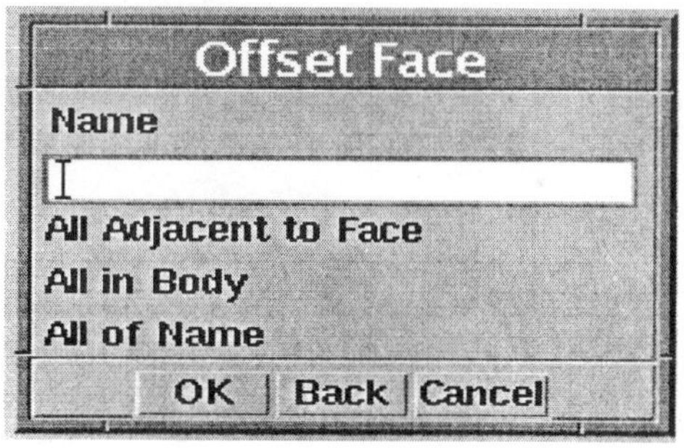

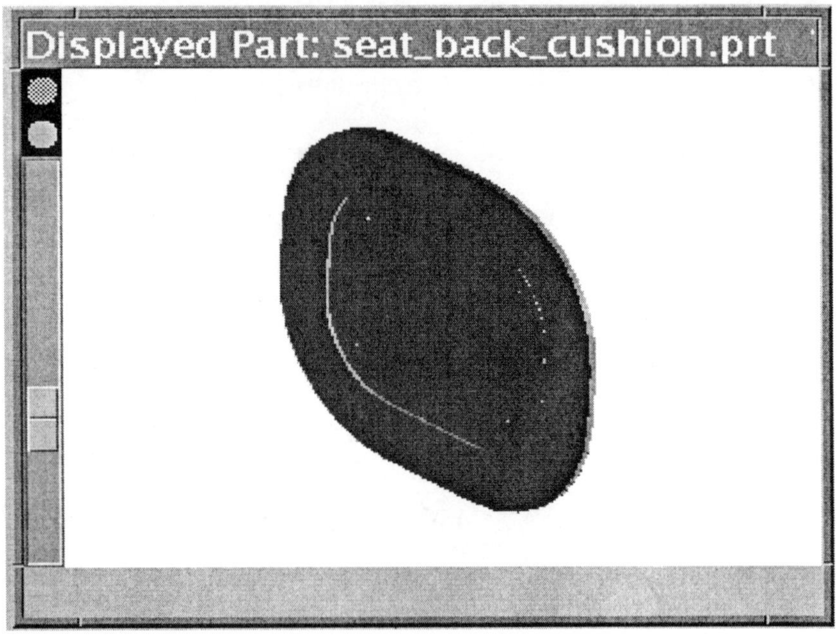

Save you work as seat_back_cushion.

Using the ANT, make the main assembly the work part. Do this by selecting seat_back_cushion with you right mouse button. Select display parent. When the assembly finishes loading, select seat_bottom_cushion and make it the work part with your right mouse button.

Go through the same steps to link the geometry from the seat bottom plate to the seat_bottom_cushion file. Don't forget the hole centers!

2.4.3 The Seat Bottom Cushion

Select Assemblies|Wave Geometry Linker. Select the region icon and choose the top face and all side faces of the bottom plate and the centers of the holes. Make sure that blank original is selected, choose OK. The boundry face is the bottom face. Choose it and select OK, twice. Thicken the sheet to .25" this time. Choose Insert|Form Feature|Thicken Sheet.

2.5 Exploded Views

2.5.1 Exploded Views

A very handy way to look at assembly files, is to create an exploded view. This is very helpful when making an assembly manual. This is a graphical way to depict how components are positioned relitive to each other in an easy to understand format. This is very helpful when your audiance is multinational. Open the file office_chair. If not already, choose Application | Assemblies

Choose Assemblies | Exploded Views | Create Explosion, the create explosion dialog appears. You can name the the explosion, we will use the default name. Select OK. You will see the word (exploded) next to the name of the view name in the graphics area. This tell you that the view is being modified. We now have to tell the system which components to move and in what direction.

Choose Assemblies | Exploded Views | Edit Explosion. The class selection dialog appears. Now we have to decied how to move components. Remember that we used the chair base as the base part and mated al other parts to it. We will now use that same part and move all other parts relitive to it. With your right mouse choose all parts above the base and select OK.

If the WCS is not lined up with the direction that you want to move the components, this need to be done now. Make sure that the WCS is oriented properly. Lets make sure that the ZC axis is pointingeither up or down in respect to the center line of the chair.

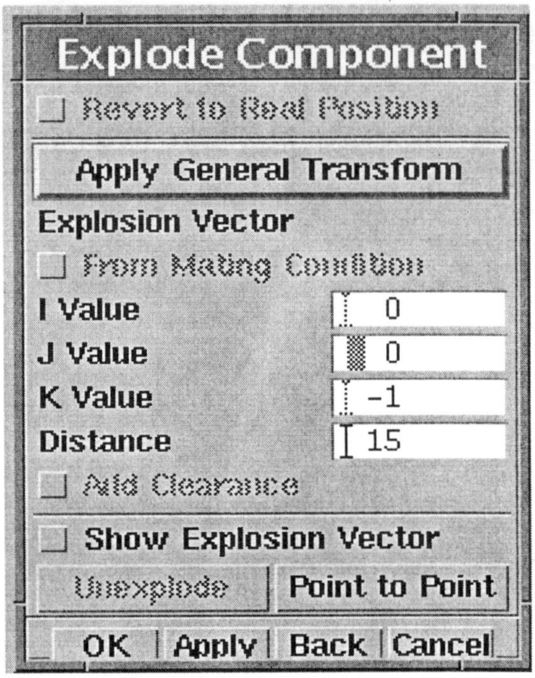

The explode component dialog box appears. We need to move all the selected components15" up. Depending on the direction of the ZC vector choose either a 1 or a -1 for the K value. We will move the components 15". Select apply.

We will now move the bottom adjustment know down 5". Choose Assemblies|Exploded Views|Edit Explosion. Select the bottom adjustment knob only. Be sure to move it in the right direction. Select Apply.

Explode the back bracket in the YC direction, away from the seat, (a positive ar negitive direction depends on the position of the WCS). Select the back bracket, select OK. The explode component dialog appears choose the proper YC direction and move the part 15". Select apply.

Next we will move the upper adjustment knob in the same direction 20". We will now seperate the back seat cushion from the back cushion plate, but first, we will move the two pieces in the back cushion together away from the seat area. First move both pieces 20" then move the back cushion 20" more. That is it for the top portion.

Now we must explode the caster area. We will only explode one caster. Be sure to select OK when you are finished or you will loose the explosion. This is what the explosion should look like. If not edit the positions of the components.

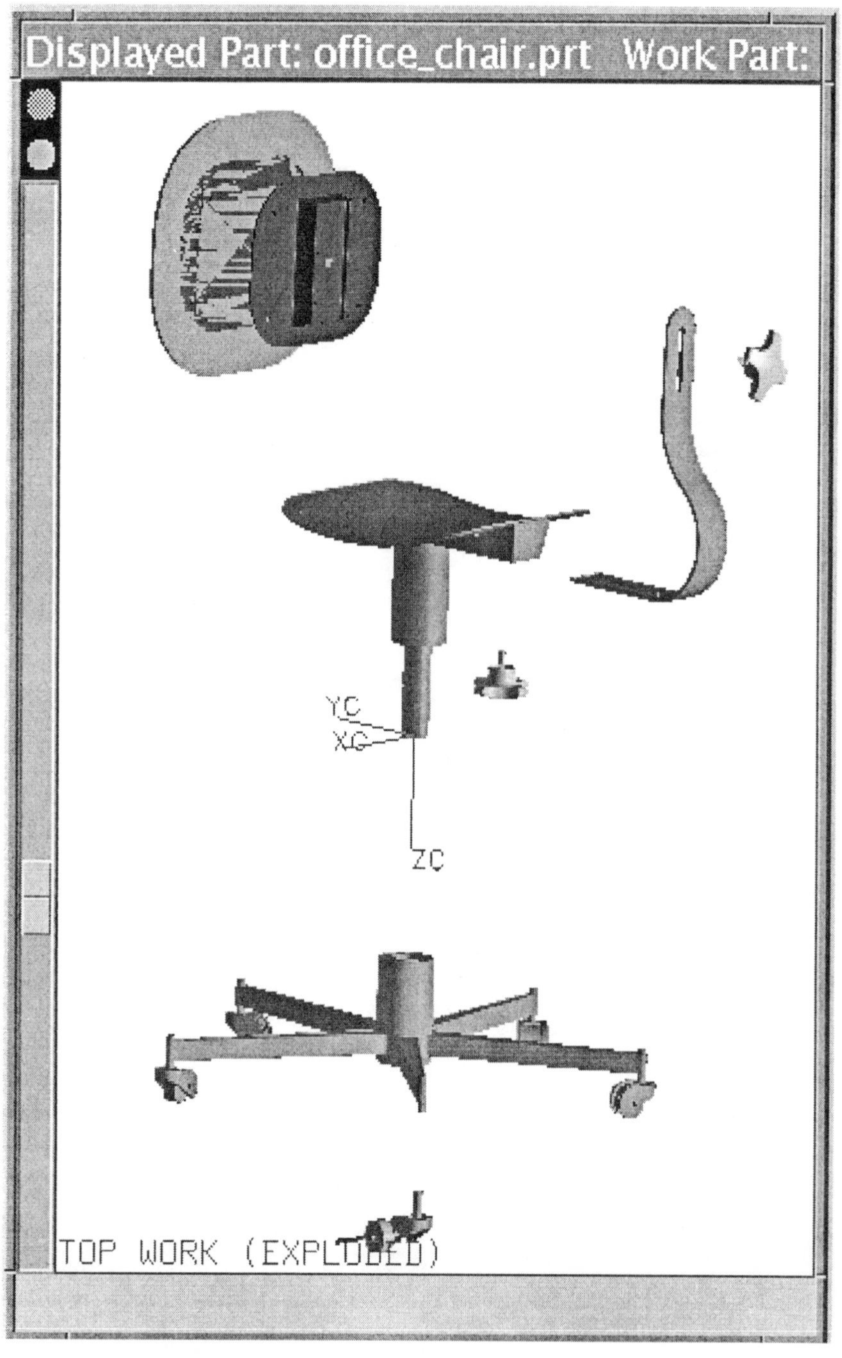

You can now togle between an exploded view and an un-exploded view, with show explosion or hide explosion.

3

Freeform Modeling

Overview

This will be an overview of the freeform functionality as related to the chair that we have been making. The freeform techniques that we will be using are very basic. A more in-depth look at freeform is reserved for an advanced Unigraphics course.

3.1 Freeform Basics

3.1.1 Unigraphics Freeform Modeling Background

Unigraphics freeform is a very powerful tool used to make shapes and objects that have curvature in more that one direction at the same time. Starting with Unigraphics version 16 there has been available the application Shape Studio. Shape Studio is used for Computer Aided Industrial Design. For the purposes of this book, all of the freeform functions that we require are also available in application modeling. We will be interested in creating sheets that can be used to trim bodies or thicken into solids. A sheet has no volumetric data. Sheets are like datum planes except that they curve.

Some suggestions for designers using freeform sheets and solids are to keep the model as simple as possible, use the minimum number of points and/or poles when creating the splines that will create the geometry.

After completing this chapter you will be able to:

- Create splines by poles and by points
- Create ruled surfaces

- Create through curves
- Create through curve meshes
- Trim bodies with sheets
- Edit spline strings by moving points and poles

3.1.2 Splines

Splines are created with the use of points or poles. When using points the spline must go through all of the points. When using poles the spline goes only through the beginning and end points and the resulting curve is approximated.

There are two types of splines, a low degree curve and a high degree curve. A low degree curve has 6 or less points/poles and a 6 point/pole curve is a 5 degree curve. The degree number is always 1 less than the number of points/poles. A high degree curve can have up to 25 points/poles, (a 24 degree curve).

Low degree curves are more flexible, they follow the poles more closely, and have improved performance. High degree curves are less flexible, may introduce unwanted oscillations, and reduce the chance of transferring data to other systems that can't handle high degree curves.

There are single or multiple segment splines. A single segment spline cannot be closed. You can create multiple segment splines with unlimited points.

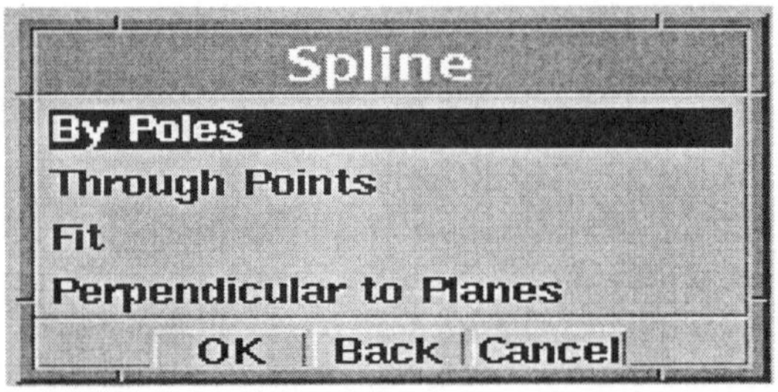

Let's create a couple of splines, one with points and one with poles. Select File|New, enter spline_1, select OK. Enter the modeling application. Choose Insert|Curve|Spline. Select by poles, OK. The spline by poles dialog box appears.

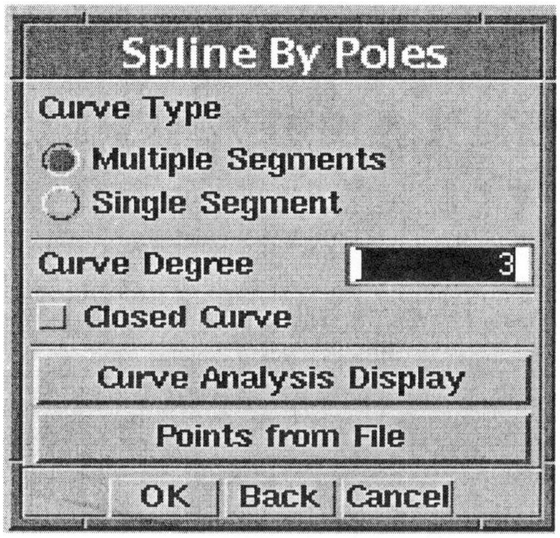

The spline by poles dialog allows you to specify multiple or single segments. A 3 degree curve is the easiest to solve. Accept these defaults and select OK. The point constructor dialog appears, it is the same dialog that you have seen before for other feature placement. To start the spline at the origin, make sure that the XC, YC and ZC coordinates are set to 0,0,0. Select OK, the first pole is placed at the origin. You can then place other points with cursor position or you can enter point values. We will use the cursor for this example. The specify points dialog appears select yes and OK. Cancel the point constructor dialog.

When you rotate the spline you can see that it is drawn on the XC-YC plane.Lets say that we are not happy with the spline shape. We will now edit the spline. Choose Edit|Curve|Parameters. The edit curve parameters dialog appears. Select the spline. Choose edit pole, make sure that the edit method is set to move pole. The cursor changes to a crosshair. Select the poles that you want to move and choose OK twice. the spline is updated. Cancel the edit curve parameters dialog.

The shape of this spline is not important in this example.

Move the WCS in the -YC direction a little, and now create a spline using the through points method. Choose Insert|Curve|Spline. Choose through points.Select OK. The spline dialog appears, choose point constructor. If needed reset back to all zeroes. Select OK. with the cursor select 4 or 5 more points in the graphics area. select OK three times. A spline is created that goes through the points specified. You can edit this spline in the same manner as you edited the first spline.

In order to use these splines to create a freeform sheet, we need to join the splines together. Let's create a new file called sheet_1. We will create a block and draw, using sketcher, a spline on each side face. The block dimensions are 6"x6"x6", placed at the origin. The block will serve as a guide to know where to put the splines. The spline end points will need to be constrained coincident with the adjacent spline.

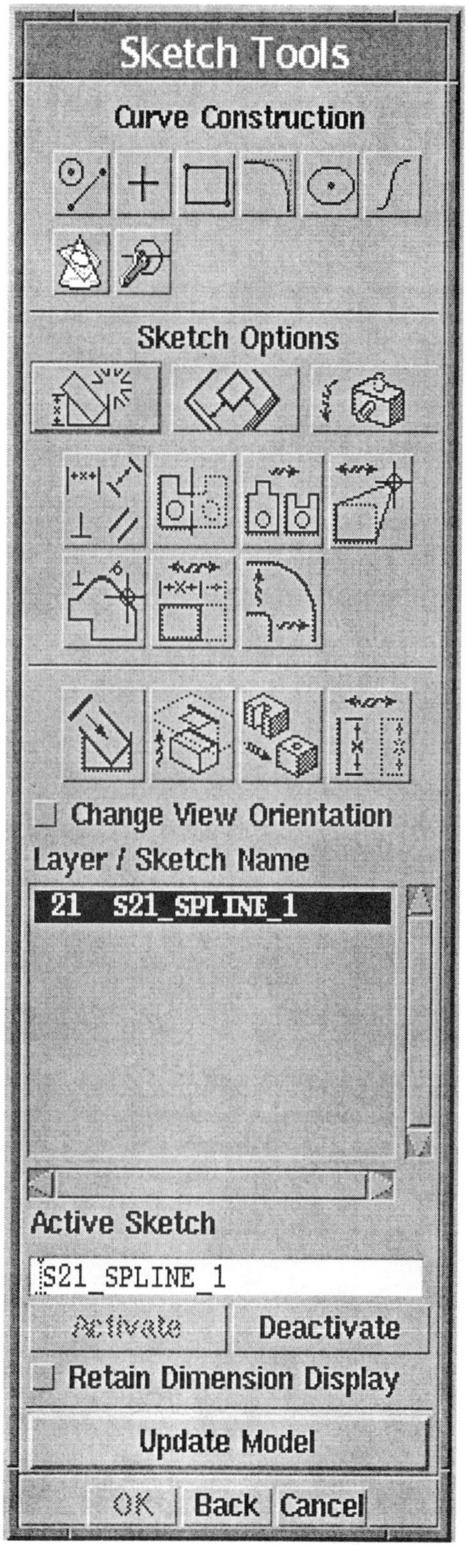

The resulting freeform sheet will be used to trim off the top of the block. Lets change the display properties of the block. Choose Edit|Object Display. Select the block, choose OK. Make the block 60% translucent, (remember to see translucency you must be in studio display). Change the line font to dashed. Select OK to accept the changes. You can now see that when you go to wireframe display the edges are dashed. Make sure that you are viewing the block in the trimetric orientation. Choose Insert|Sketch, change the work layer to 21 for the first sketch and create a sketch on the front, right plane. Use a line in the YC direction as the horizontal reference line. Choose OK. Select the spline icon. For this first spline we will choose by poles, select OK. The spline by poles dialog box appears, accept the defaults, select OK. We want to draw this first spline on the face but not on a line. If you choose a line the pole will snap to a control point of the line. choose a point near the line but do not highlight the line. We will constrain the line later. Choose 4 points on the face from one side to the other. Since the degree of curve was set to 3, we created a single segment spline with 4 points. Cancel the point constructor dialog. Now constrain the end points to point on curve.

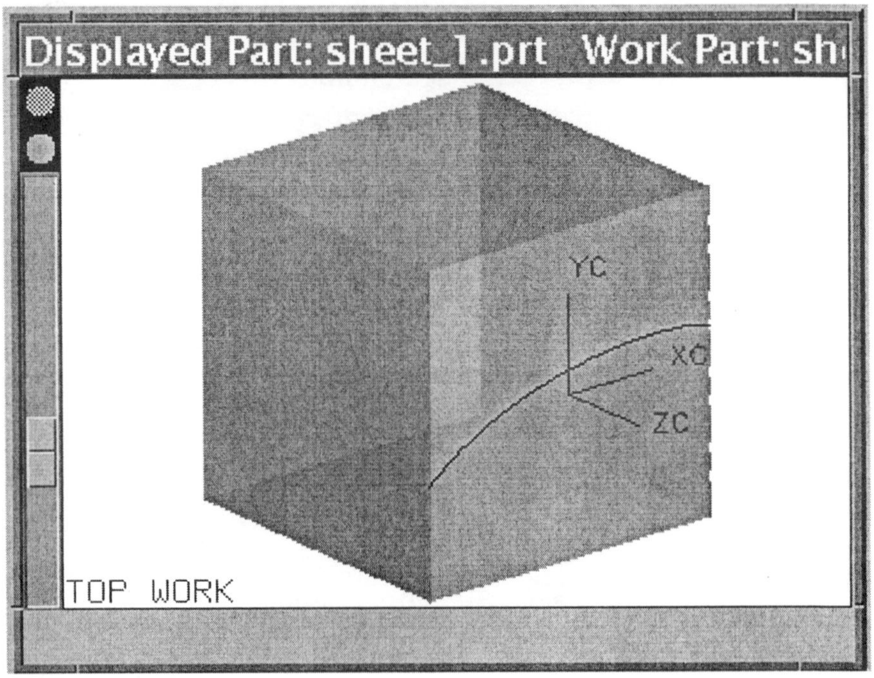

This is what the first spline will look like. The shape is not what is important, but remember the hints at the beginning of this chapter. Keep it simple with out any very sharp bends. Continue around the block use 2 splines with points and 2 with poles constrain the end points to be coincident.

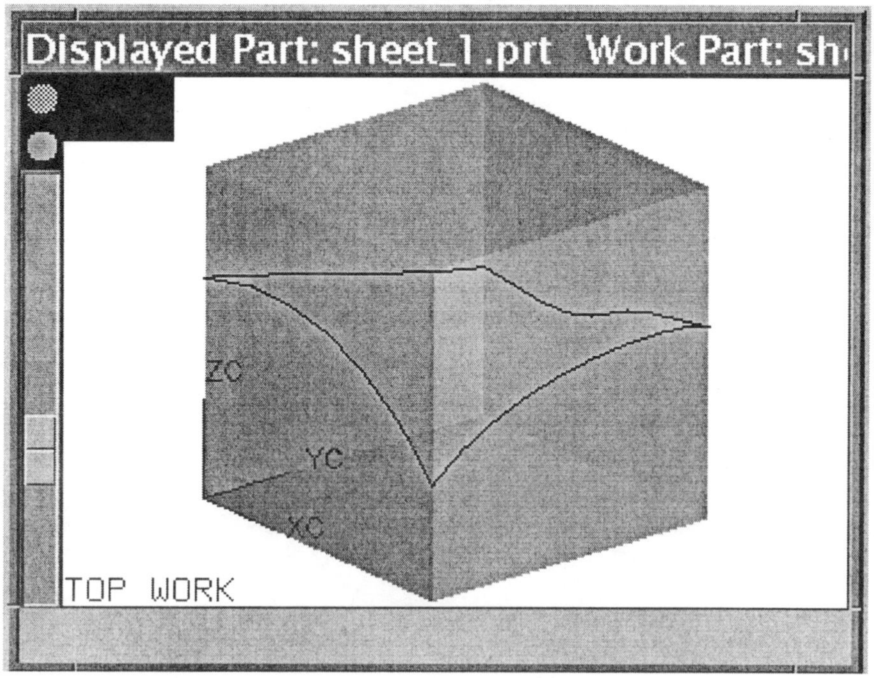

The 4 splines will look something like this on the 4 faces, connected at the end points.

We will now use this as a model to construct some different kinds of freeform sheets and/or surfaces and compare the results. To help clarify what we are doing, blank the block to see only the splines. choose Edit|Blank|Blank, select the block and select OK.

3.2 Freeform sheet types

The types of freeform shapes that we are concerned with in this book are ruled and through curves. Ruled and through curves are very similar. The main difference is that ruled allow only two section strings and are always linear between the strings. Through curves can have up to 150 strings. Also with through curve there are two types, through curve and through curve mesh. We will look at all three of these types and then apply them as we continue to model our chair. As with everything else, sheets belong on a special layer. Sheets belong on layers 81-100.

3.2.1 Ruled surfaces

Remember that ruled can use only 2 section strings. We have 4 in our model. Choose Insert|Freeform Feature|Ruled. the ruled dialog box appears. Lets start by choosing 2 adjacent splines. the place that you choose on the spline will determine the final shape of the surface. direction vector arrows are created that inform us what direction the sheet will be formed and the starting point of the sheet. In order to see what will happen with the different choices we will create this first sheet in two different ways.

Choose the spline that is on the left face first near the end at the point closest to the front. Notice the cue line. Select OK. A direction vector arrow appears informing us that the sheet will be created starting at this near corner along the spline. The cue line is now looking for the next string. Choose the next spline at the farthest end, select OK, twice.

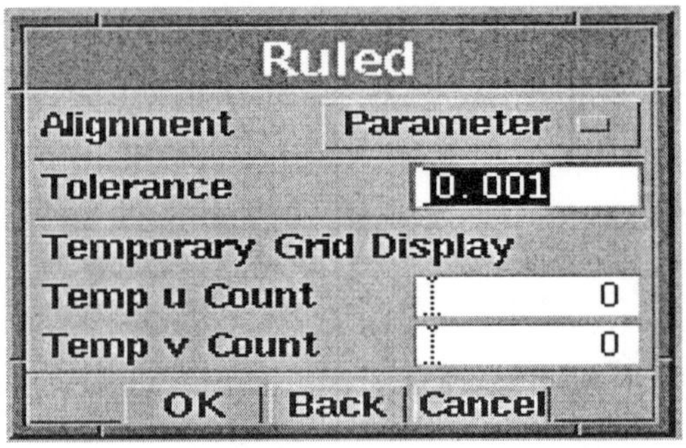

The ruled dialog box appears. The defaults are acceptable for this exercise. Select OK. Because there is a block in our model, (it is blanked), the Boolean operation dialog appears. Choose create. Observe how the sheet was formed. As you can see, the sheet is not what we want. The sheet is doubling back on itself. Select Edit|Undo List|Ruled to remove the sheet and we will do it again this time making sure that the arrows are starting from the same corners.

The new sheet is more like what we expected. Remove this sheet and choose other 2 string combinations and note the differences in the sheets.

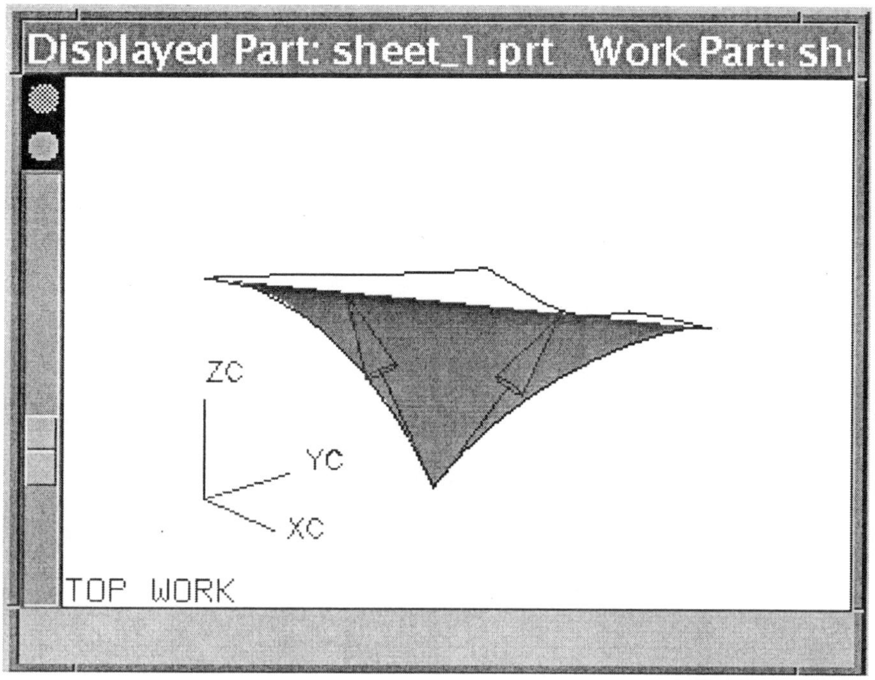

3.2.2 Through curves

The next type of sheet we will make is through curves. You can now pick more than 2 strings. In our example there are only two strings so this result is the same as ruled. Choose Insert|Freeform Feature|Through Curves. Select two opposite strings, (remember to select the strings close to the same side). the through curves dialog box appears. As you can see the box has many more choices as to how the sheet is formed. select OK, select create.

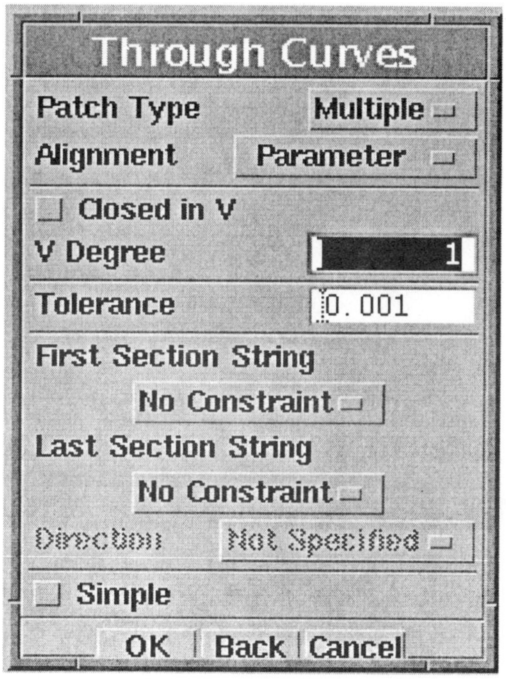

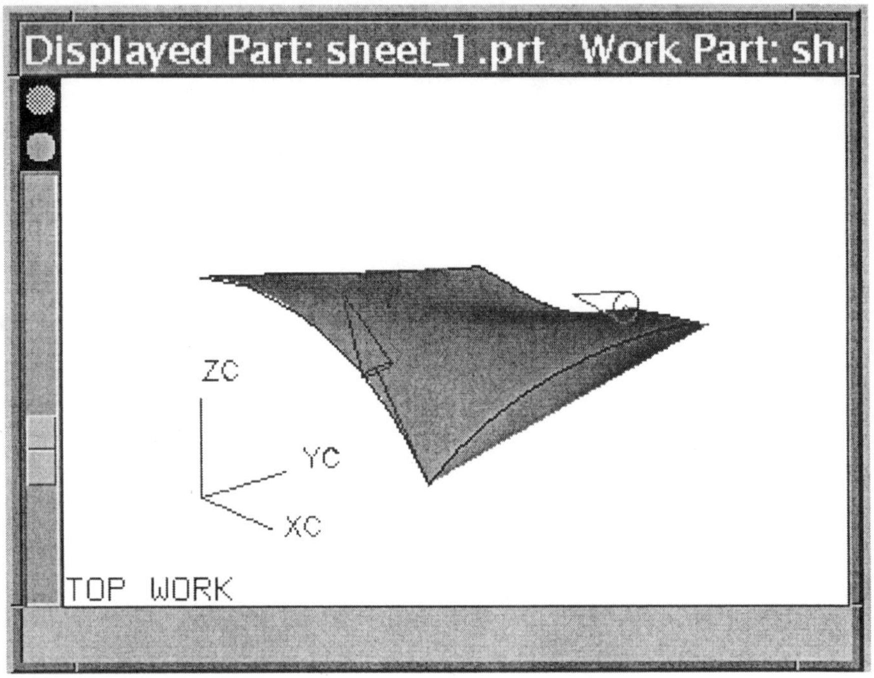

As you can see the sheet only takes into account two strings. We need to create a sheet with curvature in all directions.

3.2.3 Through Curve Mesh

The next kind of freeform sheet that we will use is through curve mesh. Undo the last sheet creation and we will build another. Choose Insert|Freeform Feature|Through Curve Mesh. Notice that the cue line is now asking you to select primary string 1. Make sure that the view is in trimetric. Select the left front spline as primary string 1, select OK. Choose the opposite spline as primary string 2, select OK. since we have only 2 primary strings select OK to inform the system that we are finished selecting primary strings. Next we must select the cross strings. The position of the cursor does not matter for cross string selection. Select the first cross and choose OK, select cross string 2 and choose OK. Since we only have 2 cross strings, select OK to get to the select spine dialog. Since we have not specified a spine select OK to get to the through curve mesh dialog.

The defaults are fine. Select OK, select create, (remember we only get this dialog because there is a blanked block in the part file). Cancel the through curve mesh dialog. Lets change the color of this sheet to yellow. Unblank the block. Switch back to layer 1. We will now trim off the top of the block with the sheet that we just created.

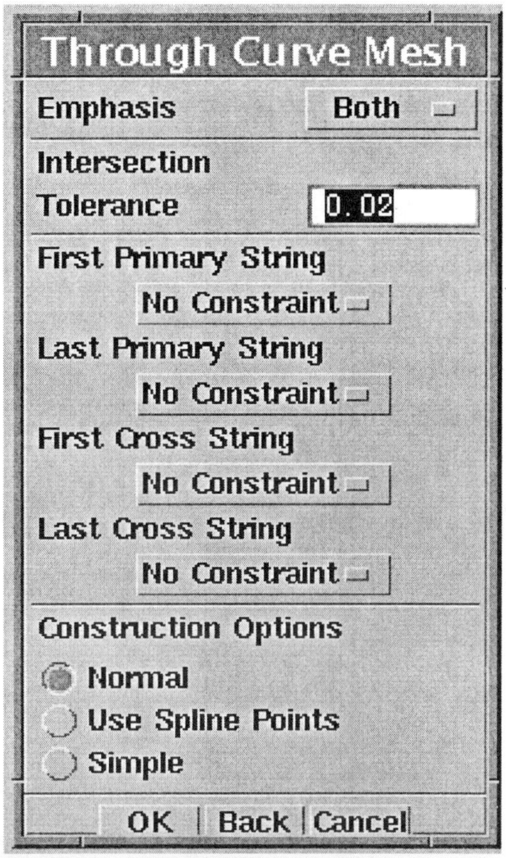

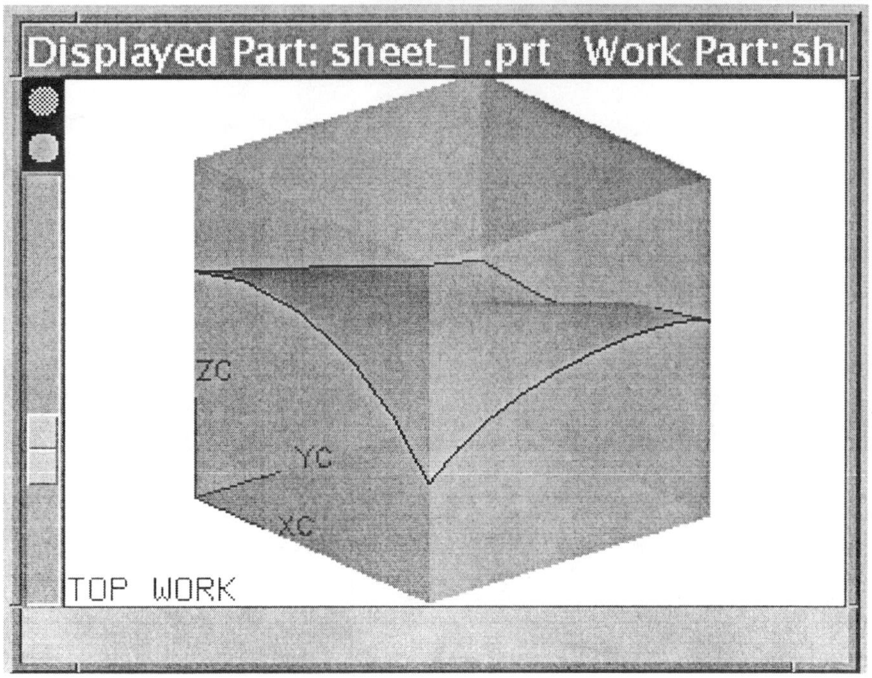

Choose Insert|Feature Operation|Trim. The target body is the block because that is what we are trimming away. Select the body, select OK. Select the sheet that you just created as the face or datum plane. Pay close attention to the direction vector arrow. The arrow points in the direction of the removed material. Make sure that the arrow is pointing to the top of the block. Cancel the trim body dialog.

As you can see the sheet and the new top of the block are at the exact same place as evidenced by the multi-colored top. Choose Format|Layer Settings, and make all layers except 1 invisible. Change the translucency back to 0% and the line font to solid, select OK.

We now have a block that is trimmed with a freeform top.

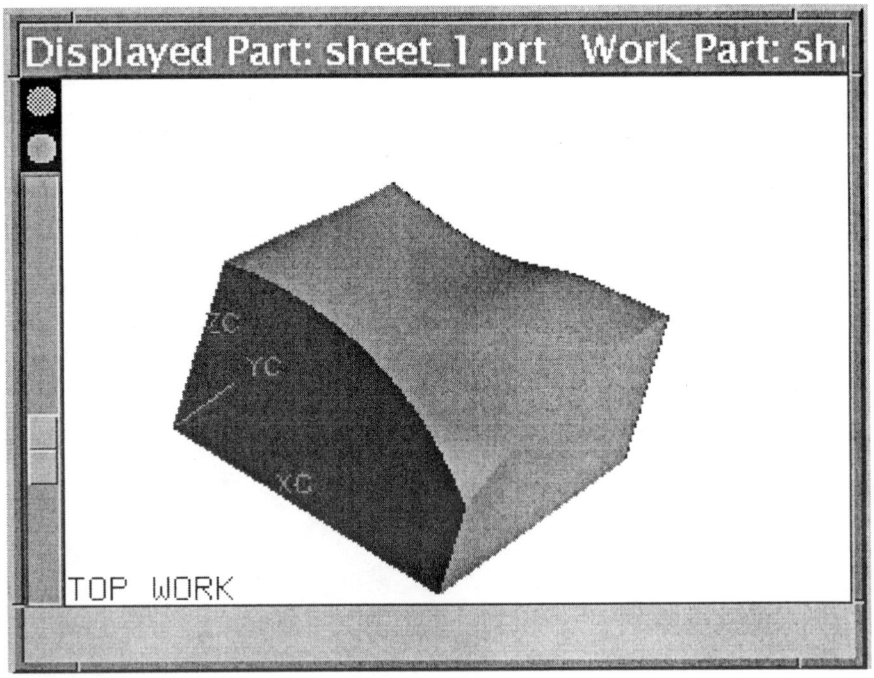

3.3 The seat cushions

3.3.1 The seat back cushion

We now have the tools necessary to model the seat cushions. We will start with the seat back cushion. make sure that the WCS is in the middle of the part at the back face.

We will construct a block around this part and trim it away with sheets. Choose Insert|Form Feature|Block. Position the block so the seat back part is in the center of the block. Choose create for the Boolean operation. Next we will trim away the sides of the block. To do this we will offset the outside edge of the seat_back1.5". Choose Insert|Curve Operation|Offset. The offset curve dialog box appears select the outside back edge and offset it a distance of 1.5". Make sure that the offset is in the proper direction. Select OK. We will now trim away the sides of the block. Assume that the block is a block of foam that we will use for the seat cushion.

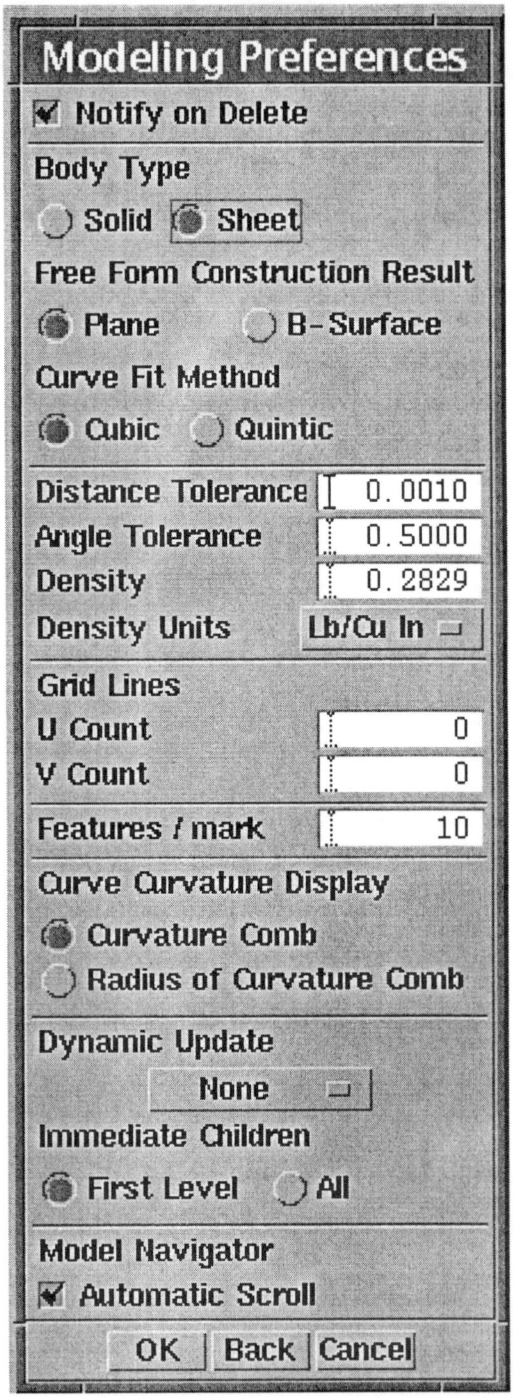

Next we will extrude the offset curves to form a sheet to trim away the outside of the block. The default states that if the section string is closed that a solid is formed. We want a sheet. Choose Preferences|Modeling, change body type from solid to sheet, select OK. Now when you extrude the section string it will form a sheet instead of a solid. Choose Insert|Form Feature|Extrude, select the offset curves, select OK, select direction and distance, make sure that the direction vector arrow is pointing into the block, choose OK. Make sure that this sheet is larger that the thickness of the block. Start at -1" and end at 5". Cancel the extruded body dialog box.

Before we trim the outside away from the block we can create some splines to form the contoured portion of the seat back cushion. If we do this now we can use the block as a guide for the splines.

If not already, open the file seat_back_cushion. We will start by putting a spline on the top face of the block. Move the WCS to the top face, a corner is fine. Make the work layer the layer that has the block and everything else visible only or invisible. We only want to apply our spline to the top face of the block. Begin with a front corner and make a spline by poles with 4 points and edit the shape to the desired shape.

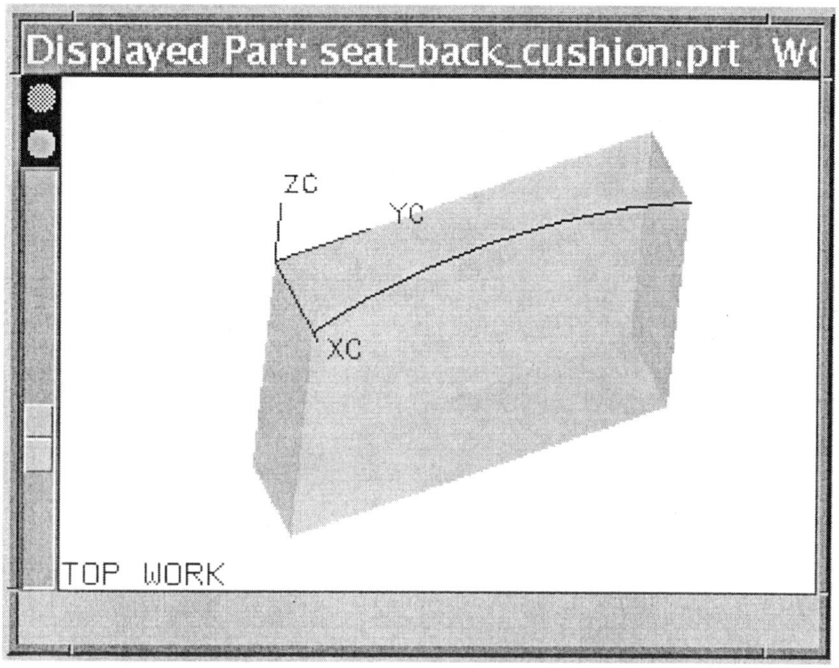

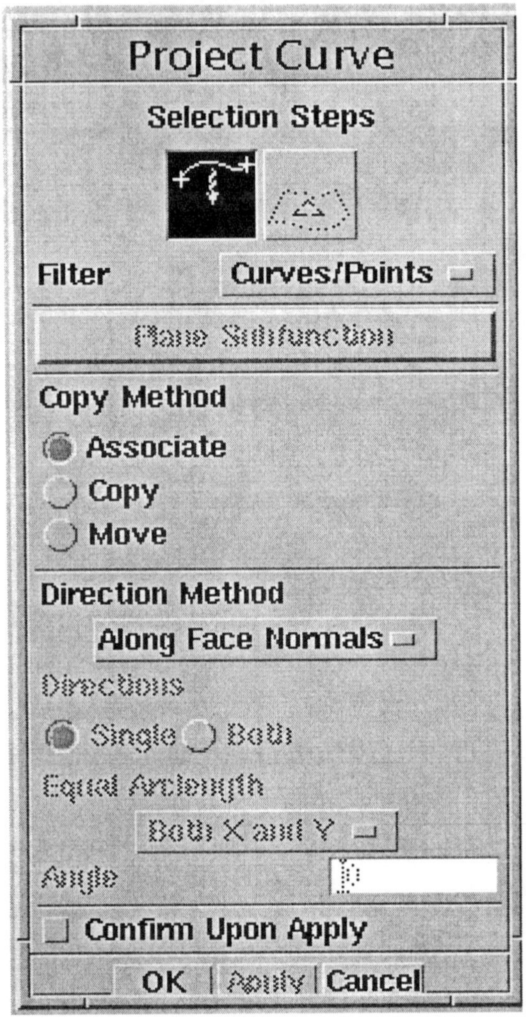

Next we need to create another spline just like the first one on the bottom face. Choose Insert|Curve Operation|Project. The project curve dialog appears. Choose the spline select OK to advance to the next selection step. Choose the bottom face of the block as the face to project the curve onto. Select OK and the curve is projected onto the face. This face is associative, meaning that if the original spline is changed the bottom one will also change.

Create a spline in the same way on the left side face and project it onto the opposite side face. Remember to move and orient the WCS to the side face.

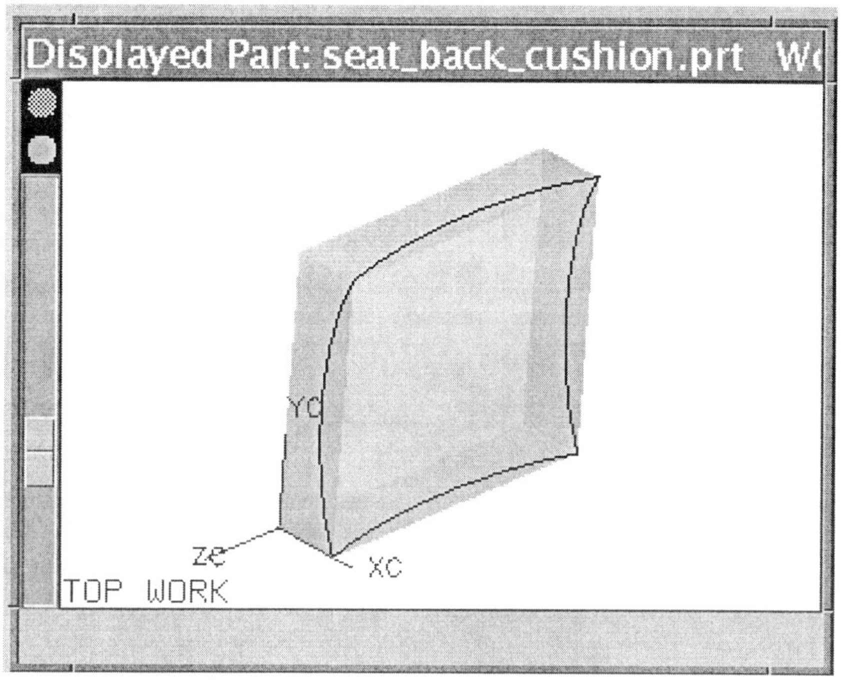

Using the 4 splines create a freeform sheet. Choose Insert|Freeform Feature|Through Curve Mesh. Select a spline near one end. Observe the direction vector arrow. Select the next primary string, making sure that the arrows point in the same direction.

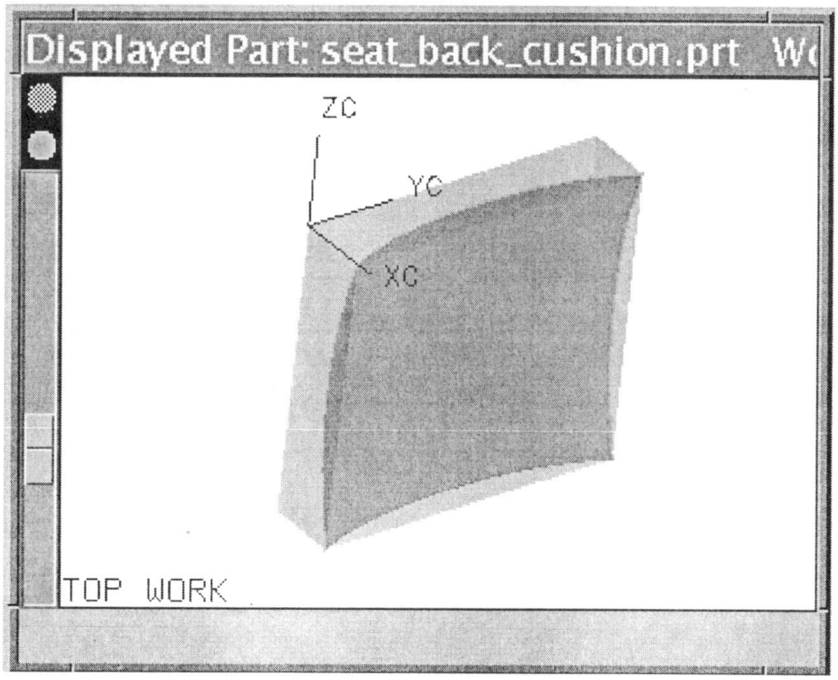

Choose Insert|Feature Operation|Trim, select the block, select OK. Select the sheet face to trim the block with. Make sure that the direction arrows are pointing away from the center of the block. If you forgot to create the sheet on layer 82, move the sheet to that layer now and make it invisible. Subtract the original seat_back_cushion shape from the block. The target solid is the block and the tool solid is the original part.

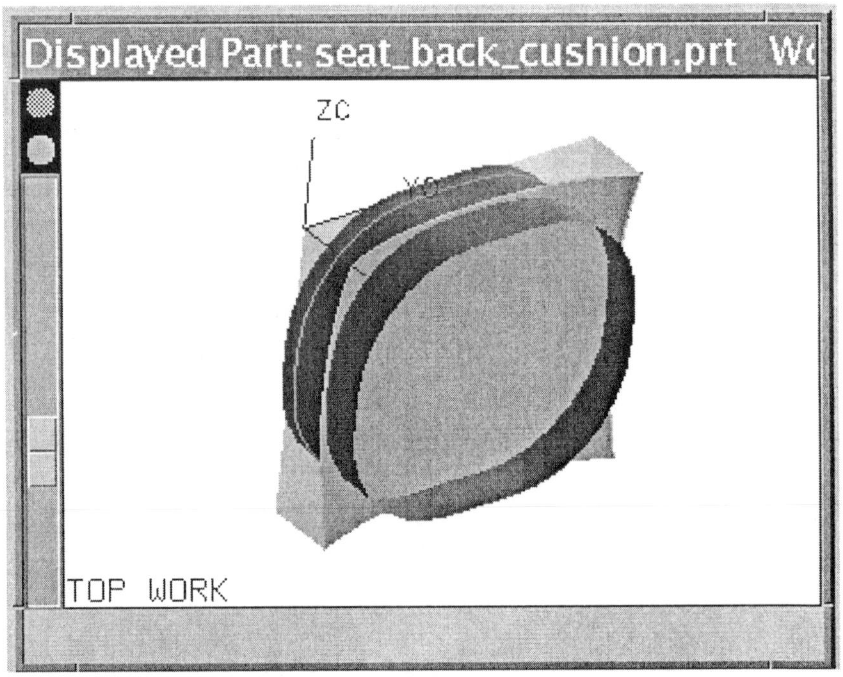

Now trim the outside of the block away with the offset curve sheet. We will now blend the edges. Start with the front edge and blend it to a 1" radius. Remember to add tangent edges. Blend the back edge to .75"

Make all layers invisible except for the block.

Use your imagination and model the seat bottom cushion following the steps that we used to make the seat back cushion. View the assembly and see the changes to the file.

Displayed Part: office_chair.prt Work Part
ZC
YC
TOP WORK

Index

Symbols

Numerics

A

B

C

D

E

F

G

H

I

L

M

N

O

P

Q

R

S

T

U

V

W

X

Notes

Notes

Notes

Notes

Notes

Notes

Notes